Lecture Notes in Computer Science 13608

More information about this series at https://link.springer.com/bookseries/558

Anthony W. Lin · Georg Zetzsche ·
Igor Potapov (Eds.)

Reachability Problems

16th International Conference, RP 2022
Kaiserslautern, Germany, October 17–21, 2022
Proceedings

Editors
Anthony W. Lin
Department of Computer Science
TU Kaiserslautern
Kaiserslautern, Germany

Max Planck Institute for Software Systems
Kaiserslautern, Germany

Igor Potapov
University of Liverpool
Liverpool, UK

Georg Zetzsche
Max Planck Institute for Software Systems
Kaiserslautern, Germany

ISSN 0302-9743   ISSN 1611-3349 (electronic)
Lecture Notes in Computer Science
ISBN 978-3-031-19134-3   ISBN 978-3-031-19135-0 (eBook)
https://doi.org/10.1007/978-3-031-19135-0

This Springer imprint is published by the registered company Springer Nature Switzerland AG
The registered company address is: Gewerbestrasse 11, 6330 Cham, Switzerland

Preface

This volume contains the papers presented at the 16th International Conference on Reachability Problems (RP 2022), organized by the Max Planck Institute for Software Systems (MPI-SWS) and the University of Kaiserslautern, Germany. Previous events in the series were located at the University of Liverpool, UK (2021), Université Paris Cité, France (2020), Université Libre de Bruxelles, Belgium (2019), Aix-Marseille University, France (2018), Royal Holloway, University of London, UK (2017), Aalborg University, Denmark (2016), the University of Warsaw, Poland (2015), the University of Oxford, UK (2014), Uppsala University, Sweden (2013), the University of Bordeaux, France (2012), the University of Genoa, Italy (2011), Masaryk University, Czech Republic (2010), École Polytechnique, France (2009), the University of Liverpool, UK (2008), and Turku University, Finland (2007).

The aim of the conference is to bring together scholars from diverse fields with a shared interest in reachability problems, and to promote the exploration of new approaches for the modeling and analysis of computational processes by combining mathematical, algorithmic, and computational techniques. Topics of interest include (but are not limited to) reachability for infinite state systems; rewriting systems; reachability analysis in counter/timed/cellular/communicating automata; Petri nets; computational game theory, computational aspects of semigroups, groups, and rings; reachability in dynamical and hybrid systems; frontiers between decidable and undecidable reachability problems; complexity and decidability aspects; predictability in iterative maps; and new computational paradigms.

We are very grateful to our invited speakers, who gave the following talks:

- **Michael Benedikt**, University of Oxford, UK:
  "The Past and Future of Embedded Finite Model Theory"
- **Laura Ciobanu**, Heriot-Watt University, UK:
  "Post's Correspondence Problem: from computer science to algebra"
- **Wojciech Czerwiński**, University of Warsaw, Poland:
  "Recent Advances on the Reachability Problem for VASSes by Examples"
- **Rupak Majumdar**, MPI-SWS, Germany:
  "Decidability Results for Context-Bounded Analysis of Systems"
- **Sharon Shoham**, Tel Aviv University, Israel:
  "SAT-Based Invariant Inference and Its Relation to Concept Learning"

The conference received 36 submissions (14 regular and 22 presentation-only submissions) from which one regular paper was withdrawn. Each submission was carefully reviewed by three Program Committee (PC) members. Based on these reviews, the PC decided to accept eight regular papers and 22 presentation-only submissions, in addition to the five invited speakers contributions. The members of the PC and the list of external reviewers can be found at the end of this preface. We are grateful for the high-quality work produced by the PC and the external reviewers. Overall this volume contains eight contributed papers and three papers from invited speakers which cover their talks. Abstracts of all invited talks can be found later in this front matter.

The conference also provided the opportunity to other young and established researchers to present work in progress or work already published elsewhere. This year in addition to the eight regular submissions, the PC accepted 22 high-quality informal presentations on various reachability aspects in theoretical computer science. A list of accepted presentation-only submissions is given later in this front matter.

So overall, the conference program consisted of five invited talks, eight presentations of contributed papers, and 22 informal presentations in the area of reachability problems, stretching from results on fundamental questions in mathematics and computer science up to efficient solutions of practical problems.

It is a pleasure to thank the team behind the EasyChair system and the Lecture Notes in Computer Science team at Springer, who together made the production of this volume possible in time for the conference. Finally, we thank all the authors and invited speakers for their high-quality contributions, and the participants for making RP 2022 a success. We are also very grateful to Alfred Hofmann for the continuous support of the event in the last decade and to Ronan Nugent for the supporting this year's conference, and also to the London Mathematical Society and Springer for their financial sponsorship.

October 2022

Anthony W. Lin
Georg Zetzsche
Igor Potapov

Organization

Program Committee

C. Aiswarya	Chennai Mathematical Institute, India
Mohamed Faouzi Atig	Uppsala University, Sweden
Udi Boker	Reichman University, Israel
Benedikt Bollig	CNRS, LMF, ENS Paris-Saclay, France
Michaël Cadilhac	DePaul University, USA
Yu-Fang Chen	Academia Sinica, Taiwan
Stéphane Demri	CNRS, LMF, ENS Paris-Saclay, France
Moses Ganardi	MPI-SWS, Germany
Julian Gutierrez	Monash University, Australia
Christoph Haase	University of Oxford, UK
Peter Habermehl	IRIF, Université Paris Cité, France
Piotr Hofman	University of Warsaw, Poland
Naoki Kobayashi	University of Tokyo, Japan
Dietrich Kuske	TU Ilmenau, Germany
Jérôme Leroux	CNRS, LaBRI, Université de Bordeaux, France
Anthony W. Lin (Co-chair)	University of Kaiserslautern, Germany
K. Narayan Kumar	Chennai Mathematical Institute, India
Guillermo Perez	University of Antwerp, Belgium
Igor Potapov (Co-chair)	University of Liverpool, UK
Gabriele Puppis	Udine University, Italy
Karin Quaas	University of Leipzig, Germany
Krishna S.	IIT Bombay, India
Mahsa Shirmohammadi	CNRS, IRIF, Université Paris Cité, France
Daniel Stan	University of Kaiserslautern, Germany
Alwen Tiu	Australian National University, Australia
Zhilin Wu	Chinese Academy of Sciences, China
Georg Zetzsche (Co-chair)	MPI-SWS, Germany

Additional Reviewers

Simon Jantsch
Fribourg Laurent
Julia Padberg
Amaury Pouly

Abstracts of Invited Talks

The Past and Future of Embedded Finite Model Theory

Michael Benedikt

University of Oxford

Abstract. Embedded finite model theory refers to a formalism for describing finite structures over an uninterpreted signature, which sit within an infinite interpreted structures. Some theory was developed in the 1990's and early 2000's, with a focus on the real field. But the theory applies to arbitrary theories, and is relevant to recent development on graph querying and analysis of data-driven programs involving arithmetic.

In this invited paper we review the framework and some of the basic results on it. We also discuss some open questions, along with some work in progress, joint with Ehud Hrushovski.

Post's Correspondence Problem: From Computer Science to Algebra

Laura Ciobanu (iD)

Heriot-Watt University and Maxwell Institute, Edinburgh EH14 4AS, Scotland
L.Ciobanu@hw.ac.uk

Abstract. In this short survey we describe recent advances on the Post Correspondence Problem in group theory that were inspired by results in computer science. These algebraic advances can, in return, provide a source of interesting problems in more applied, computational settings.

Post's Correspondence Problem (PCP) is a classical decision problem in theoretical computer science that asks whether for a pair of free monoid morphisms $g, h : \Sigma^* \to \Delta^*$ there is any $x \in \Sigma^*$ such that $g(x) = h(x)$. One can similarly phrase a PCP for general groups, rather than free monoids, by asking whether pairs g, h of group homomorphisms agree on any inputs. This leads to interesting and unexpected (un)decidability results for PCP in groups.

Keywords: Post correspondence problem · Free and hyperbolic groups · Free monoids · Nilpotent groups · Decidability

Recent Advances on the Reachability Problem for VASSes by Examples

Wojciech Czerwiński

University of Warsaw
wczerwin@mimuw.edu.pl

Abstract. I will briefly describe recent advances on understanding the complexity of the reachability problem for vector addition systems (or equivalently for vector addition systems with states - VASSes). I plan to present a few involved VASS examples in small dimensions, which illustrate various aspects of hardness of VASSes and various techniques of proving lower complexity bounds. If time allows I will also briefly discuss VASSes with a stack.

Supported by the ERC grant INFSYS, agreement no. 950398.

Decidability Results for Context-Bounded Analysis of Systems

Rupak Majumdar

MPI-SWS, Kaiserslautern, Germany
`rupak@mpi-sws.org`

Abstract. Automated analysis of multithreaded shared memory programs is a core problem in verification. While the general verification problem for these systems is undecidable, already when there are only two recursive threads, there has been a lot of work to find appropriate underapproximations. Context-bounding is one such underapproximation technique. In context bounded analysis, we set an a priori bound K and restrict attention to only those runs of the system in which each thread is interrupted at most K times. It turns out that many verification problems become decidable under the restriction of context bounding. In this talk, I will provide a survey of recent results in this area. Specifically, we shall consider context-bounded safety and liveness verification for systems in which threads can spawn new threads, as well as practically-motivated restrictions of the problem such as thread pooling. (Joint work with Pascal Baumann, Moses Ganardi, Ramanathan Thinniyam, and Georg Zetzsche.)

SAT-Based Invariant Inference and Its Relation to Concept Learning

Yotam M. Y. Feldman and Sharon Shoham

Tel Aviv University, Israel

Abstract. This paper surveys results that establish formal connections and distinctions between SAT-based invariant inference and exact concept learning with queries, showing that learning techniques and algorithms can clarify foundational questions, illuminate existing algorithms, and suggest new directions for efficient invariant inference.

Presentation-Only Submissions

The Pseudo-Reachability Problem for Affine Dynamical Systems

Julian D'Costa, Toghrul Karimov, Rupak Majumdar, Joël Ouaknine,
Mahmoud Salamati, and James Worrell

Abstract. We study fundamental reachability problems on pseudo-orbits
of linear dynamical systems. Pseudo-orbits can be viewed as a model
of computation with limited precision and pseudo-reachability can be
thought of as a robust version of classical reachability. Using an approach
based on o-minimality of $\mathbb{R}_{exp}$ we prove decidability of the discrete-
time pseudo-reachability problem with arbitrary semialgebraic targets for
diagonalisable linear dynamical systems. We also show that our method
can be used to reduce the continuous-time pseudo-reachability problem
to the (classical) time-bounded reachability problem, which is known to
be conditionally decidable. In short, we show how to use logical methods
(in contrast to the usual number theory) to decide versions of the classical
reachability problem for linear dynamical systems.

Pairwise Reachability Oracles and Preservers Under Failures

Diptarka Chakraborty, Kushagra Chatterjee, and Keerti Choudhary

Abstract. In this paper, we consider reachability oracles and reachability preservers for directed graphs/networks prone to edge/node failures. Let $G = (V, E)$ be a directed graph on n-nodes, and $\mathcal{P} \subseteq V \times V$ be a set of vertex pairs in G. We present the first non-trivial constructions of single and dual fault-tolerant pairwise reachability oracle with constant query time. Furthermore, we provide extremal bounds for sparse fault-tolerant reachability preservers, resilient to two or more failures. Prior to this work, such oracles and reachability preservers were widely studied for the special scenario of single-source and all-pairs settings. However, for the scenario of arbitrary pairs, no prior (non-trivial) results were known for dual (or more) failures, except those implied from the single-source setting. One of the main questions is whether it is possible to beat the $O(n|\mathcal{P}|)$ size bound (derived from the single-source setting) for reachability oracle and preserver for dual failures (or $O(2^k n|\mathcal{P}|)$ bound for k failures). We answer this question affirmatively. Below we summarize our contributions.

- For an n-vertex directed graph $G = (V, E)$ and $\mathcal{P} \subseteq V \times V$, we present a construction of $O(n\sqrt{|\mathcal{P}|})$ sized dual fault-tolerant pairwise reachability oracle with constant query time. We further provide a matching (up to the word size) lower bound of $\Omega(n\sqrt{|\mathcal{P}|})$ on the size (in bits) of the oracle for the dual fault setting, thereby proving that our oracle is (near-)optimal.
- Next, we provide a construction of $O(n + \min\{|\mathcal{P}|\sqrt{n}, \, n\sqrt{|\mathcal{P}|}\})$ sized oracle with $O(1)$ query time, resilient to single node/edge failure. In particular, for $|\mathcal{P}|$ bounded by $O(\sqrt{n})$ this yields an oracle of just $O(n)$ size. We complement the upper bound with a lower bound of $\Omega(n^{2/3}|\mathcal{P}|^{1/2})$ (in bits), refuting the possibility of a linear-sized oracle for $\mathcal{P}$ of size $\omega(n^{2/3})$.
- We also present a construction of $O(n^{4/3}|\mathcal{P}|^{1/3})$ sized pairwise reachability preservers resilient to dual edge/vertex failures. Previously, such preservers were known to exist only under single failure and had $O(n + \min\{|\mathcal{P}|\sqrt{n}, \, n\sqrt{|\mathcal{P}|}\})$ size [Chakraborty and Choudhary, ICALP'20]. We also show a lower bound of $\Omega(n\sqrt{|\mathcal{P}|})$ edges on the size of dual fault-tolerant reachability preservers, thereby providing a sharp gap between single and dual fault-tolerant reachability preservers for $|\mathcal{P}| = o(n)$.

– Finally, we provide a generic pairwise reachability preserver construction that provides a $o(2^k n|\mathcal{P}|)$ sized subgraph resilient to k failures, for any $k \geq 1$. Before this work, we only knew of an $O(2^k n|\mathcal{P}|)$ bound implied from the single-source setting [Baswana, Choudhary, and Roditty, STOC'16].

On the Identity Problem for Unitriangular Matrices of Dimension Four

Ruiwen Dong

Abstract. We show that the Identity Problem is decidable in polynomial time for finitely generated sub-semigroups of the group UT(4, $\mathbb{Z}$) of 4 × 4 unitriangular integer matrices. As a byproduct of our proof, we also show the polynomial-time decidability of several subset reachability problems in UT(4, $\mathbb{Z}$).

On the Undecidability of Loop Analysis

Laura Kovács and Anton Varonka

Abstract. Our work addresses two central questions of program analysis, termination and invariant generation of loops. Already simple infinite-state systems, such as loops with assignments only, have intrinsically complex reachable sets and render variants of these problems undecidable.

In the talk, we give an account of the existing body of work on the boundary of (un-)decidability. We contribute to the line of work related to the question raised by Braverman in 2006: "How much non-determinism can be introduced in a linear loop before termination becomes undecidable?" We show that termination of loops with a purely non-deterministic choice between linear updates cannot be answered by an algorithm. To our knowledge, this is the most restricted setting in which undecidability has been shown. Moreover, it contrasts the case of a single linear update where termination (i.e., reachability) is known decidable.

We also turn to the problem of computing the strongest algebraic invariant of a program, that is, all polynomial relations among program variables. Despite a complete algorithm for multi-path affine programs, allowing arbitrary polynomial assignments is known to result in the unsolvability of invariant generation. We point out that negative results do not actually exploit general polynomial updates. There exists no algorithm computing strongest algebraic invariants already for programs with quadratic updates or updates guarded by affine equalities.

Skolem Meets Schanuel

Yuri Bilu, Florian Luca, Joris Nieuwveld, Joël Ouaknine,
David Purser, and James Worrell

Abstract. The celebrated Skolem-Mahler-Lech Theorem states that the
set of zeros of a linear recurrence sequence is the union of a finite set and
finitely many arithmetic progressions. The corresponding computational
question, the Skolem Problem, asks to determine whether a given linear
recurrence sequence has a zero term. Although the Skolem-Mahler-Lech
Theorem is almost 90 years old, decidability of the Skolem Problem
remains open. The main contribution of this paper is an algorithm to
solve the Skolem Problem for simple linear recurrence sequences (those
with simple characteristic roots). Whenever the algorithm terminates, it
produces a stand-alone certificate that its output is correct – a set of zeros
together with a collection of witnesses that no further zeros exist. We
give a proof that the algorithm always terminates assuming two classical
number-theoretic conjectures: the Skolem Conjecture (also known as the
Exponential Local-Global Principle) and the p-adic Schanuel Conjecture.
Preliminary experiments with an implementation of this algorithm within
the tool SKOLEM point to the practical applicability of this method.

Subsequences with Gap Constraints: Complexity Bounds for Matching and Analysis Problems

Joel Day, Maria Kosche, Florin Manea, and Markus L. Schmid

Abstract. We consider subsequences with gap constraints, i.,e., length-k subsequences p that can be embedded into a string w such that the induced gaps (i.,e., the factors of w between the positions to which p is mapped to) satisfy given gap constraints $gc = (C_1, C_2, \ldots, C_{k-1})$; we call p a gc-subsequence of w. In the case where the gap constraints gc are defined by lower and upper length bounds $C_i = (L_i^-, L_i^+) \in \mathbb{N}^2$ and/or regular languages $C_i \in \mathrm{REG}$, we prove tight (conditional on the orthogonal vectors (OV) hypothesis) complexity bounds for checking whether a given p is a gc-subsequence of a string w. We also consider the whole set of all gc-subsequences of a string, and investigate the complexity of the universality, equivalence and containment problems for these sets of gc-subsequences.

A Dynamic Programming Algorithm for a Maximum
s-Clique Set on Trees

José Alberto Fernández-Zepeda, Alejandro Flores Lamas,
Matthew Hague, and Joel Antonio Trejo-Sánchez

Abstract. Given an undirected graph $G = (V_G, E_G)$, a clique C is a complete subgraph of G. In social networks analysis, the unit distance of the clique makes it challenging to model certain social concepts. For those applications, a relaxed variant of the clique is more appropriate. A s-clique Q is a maximal subgraph of G, such that the distance in G between any pair of vertices of Q is less than or equal to some positive integer s. In this sense, a clique is also a 1-clique. The maximum s-clique problem consists of finding a s-clique with the greatest amount of vertices in G. Such a problem is NP-hard for arbitrary graphs and any s. In this work, we propose a dynamic programming algorithm that solves this problem on a tree of order n in $O(s \cdot n)$ time, for $s \geq 2$. Our algorithm improves, theoretically and experimentally, the performance of previous algorithms that compute a maximum s-clique on trees.

Higher-Order Nonemptiness Step by Step

Paweł Parys

Abstract. We show a new simple algorithm that checks whether a given higher-order grammar generates a nonempty language of trees. The algorithm amounts to a procedure that transforms a grammar of order n to a grammar of order n − 1, preserving nonemptiness, and increasing the size only exponentially. After repeating the procedure n times, we obtain a grammar of order 0, whose nonemptiness can be easily checked. Since the size grows exponentially at each step, the overall complexity is n-EXPTIME, which is known to be optimal. More precisely, the transformation (and hence the whole algorithm) is linear in the size of the grammar, assuming that the arity of employed nonterminals is bounded by a constant.

The same algorithm allows to check whether an infinite tree generated by a higher-order recursion scheme is accepted by an alternating reachability automaton, because this question can be reduced to the nonemptiness problem by taking a product of the recursion scheme with the automaton. Moreover, thanks to a well-known equivalence between higher-order grammars and collapsible pushdown automata, we also obtain an algorithm for reachability in alternating collapsible pushdown automata.

A proof of correctness of the algorithm is formalised in the proof assistant Coq. Our transformation is motivated by a similar transformation of Asada and Kobayashi (2020) changing a word grammar of order n to a tree grammar of order n − 1. The step-by-step approach can be opposed to previous algorithms solving the nonemptiness problem "in one step", being compulsorily more complicated.

A Universal Skolem Set of Positive Lower Density

Florian Luca, Joël Ouaknine, and James Worrell

Abstract. The Skolem Problem asks to determine whether a given integer linear recurrence sequence (LRS) has a zero term. Decidability of this problem has been open for many decades, with little progress since the 1980s. Recently, a new approach was initiated via the notion of a Skolem set – a set of positive integers relative to which the Skolem Problem is decidable. More precisely, S is a Skolem set for a class L of integer LRS if there is an effective procedure that, given an LRS in L, decides whether the sequence has a zero in S. A recent work exhibited a Skolem set for the class of all LRS that, while infinite, had density zero. In the present paper we construct a Skolem set of positive lower density for the class of simple LRS.

On the Computation of the Algebraic Closure of Finitely Generated Groups of Matrices

Amaury Pouly, Klara Nosan, Mahsa Shirmohammadi,
James Worrell, and Sylvain Schmitz

Abstract. We investigate the complexity of computing the Zariski closure of a finitely generated group of matrices. The Zariski closure was previously shown to be computable by Derksen, Jeandel, and Koiran, but the termination argument for their algorithm appears not to yield any complexity bound. In this paper we follow a different approach and obtain a bound on the degree of the polynomials that define the closure. Our bound shows that the closure can be computed in elementary time. We also obtain upper bounds on the length of chains of linear algebraic groups.

Matching Patterns with Variables Under Edit Distance

Paweł Gawrychowski, Florin Manea, and Stefan Siemer

Abstract. A pattern α is a string of variables and terminal letters. We say that α matches a word w, consisting only of terminal letters, if w can be obtained by replacing the variables of α by terminal words. The matching problem, i.e., deciding whether a given pattern matches a given word, was heavily investigated: it is NP-complete in general, but can be solved efficiently for classes of patterns with restricted structure. If we are interested in what is the minimum Hamming distance between w and any word u obtained by replacing the variables of α by terminal words (so matching under Hamming distance), one can devise efficient algorithms and matching conditional lower bounds for the class of regular patterns (in which no variable occurs twice), as well as for classes of patterns where we allow unbounded repetitions of variables, but restrict the structure of the pattern, i.e., the way the occurrences of different variables can be interleaved. Moreover, for matching under Hamming distance, if a variable occurs more than once and its occurrences can be interleaved arbitrarily with those of other variables, even if each of these occurs just once, the problem is intractable. In this paper, we consider the problem of matching patterns with variables under edit distance. We still obtain efficient algorithms and matching conditional lower bounds for the class of regular patterns, but show that the problem becomes, in this case, intractable already for unary patterns, containing repeated occurrences of a single variable interleaved with terminals.

The Variance-Penalized Stochastic Shortest Path Problem

Jakob Piribauer, Ocan Sankur, and Christel Baier

Abstract. The stochastic shortest path problem (SSPP) asks to resolve the non-deterministic choices in a Markov decision process (MDP) such that the expected accumulated weight before reaching a target state is maximized. This paper addresses the optimization of the variance-penalized expectation (VPE) of the accumulated weight, which is a variant of the SSPP in which a multiple of the variance of accumulated weights is incurred as a penalty. It is shown that the optimal VPE in MDPs with non-negative weights as well as an optimal deterministic finite-memory scheduler can be computed in exponential space. The threshold problem whether the maximal VPE exceeds a given rational is shown to be EXPTIME-hard and to lie in NEXPTIME. Furthermore, a result of interest in its own right obtained on the way is that a variance-minimal scheduler among all expectation-optimal schedulers can be computed in polynomial time. This paper has been published at ICALP 2022.

Parameterized Safety Verification of Round-Based Shared-Memory Systems

Nathalie Bertrand, Nicolas Markey, Ocan Sankur and Nicolas Waldburger

Abstract. We consider the parameterized verification problem for distributed algorithms where the goal is to develop techniques to prove the correctness of a given algorithm regardless of the number of participating processes. Motivated by an asynchronous binary consensus algorithm [Aspnes02], we consider round-based distributed algorithms communicating with shared memory. A particular challenge in these systems is that 1) the number of processes is unbounded, and, more importantly, 2) there is a fresh set of registers at each round. A verification algorithm thus needs to manage both sources of infinity. In this setting, we prove that the safety verification problem, which consists in deciding whether all possible executions avoid a given error state, is PSPACE-complete. For negative instances of the safety verification problem, we also provide exponential lower and upper bounds on the minimal number of processes needed for an error execution and on the minimal round on which the error state can be covered.

Program Specialization as a Tool for Solving Word Equations

Antonina Nepeivoda

Abstract. The paper was presented at VPT'2021. The paper focuses on the automatic generating of the witnesses for the word equation satisfiability problem by means of specializing an interpreter WeqInt(s, E), which tests whether a substitution s of variables of a given word equation system E produces its solution. We specialize such an interpreter w.r.t. E, while s is unknown. We show that several variants of such interpreters, when specialized using the basic unfold/fold specialization methods, are able to decide the satisfiability problem for some sets of the word equations whose left- and right-hand sides share variables. We prove that the specialization process w.r.t. the constructed interpreters is sound, i.e. gives a simple syntactic criterion of the satisfiability, and compare the results of the suggested approach with the results produced by some known SMT-solvers.

Distributed Controller Synthesis for Deadlock Avoidance

Hugo Gimbert, Corto Mascle, Anca Muscholl, and Igor Walukiewicz

Abstract. We consider the distributed control synthesis problem for systems with locks. The goal is to find local controllers so that the global system does not deadlock. With no restriction this problem is undecidable even for three processes each using a fixed number of locks. We propose two restrictions that make distributed control decidable. The first one is to allow each process to use at most two locks. The problem then becomes complete for the second level of the polynomial time hierarchy, and even in Ptime under some additional assumptions. The dining philosophers problem satisfies these assumptions. The second restriction is a nested usage of locks. In this case the synthesis problem is Nexptime-complete. The drinking philosophers problem falls in this case.

The Membership Problem for Hypergeometric Sequences with Rational Parameters

Amaury Pouly, Klara Nosan, Mahsa Shirmohammadi, and James Worrell

Abstract. We investigate the Membership Problem for hypergeometric sequences: given a hypergeometric sequence $\langle u_n \rangle_{n=0}^{\infty}$ of rational numbers and a target $t \in \mathbb{Q}$, decide whether t occurs in the sequence. We show decidability of this problem under the assumption that in the defining recurrence $p(n)u_{n+1} = q(n)u_n$, the roots of the polynomials $p(x)$ and $q(x)$ are all rational numbers. Our proof relies on bounds on the density of primes in arithmetic progressions. We also observe a relationship between the decidability of the Membership problem (and variants) and the Rohrlich-Lang conjecture in transcendence theory.

On the Expressive Power of String Constraints

Joel Day, Vijay Ganesh, Nathan Grewal, and Florin Manea

Abstract. We investigate properties of strings which are expressible by canonical types of string constraints. Specifically, we consider a landscape of 20 logical theories, whose syntax is built around combinations of four common elements of string constraints: language membership (e.g. for regular languages), concatenation, equality between string terms, and equality between string-lengths. For a variable x and formula f from a given theory, we consider the set of values for which x may be substituted as part of a satisfying assignment, or in other words, the property f expresses through x. Since we consider string-based logics, this set is a formal language. We firstly consider the relative expressive power of different combinations of string constraints by comparing the classes of languages expressible in the corresponding theories, and are able to establish a mostly complete picture in this regard. Secondly, we consider the question of deciding whether the language or property expressed by a variable/formula in one theory can be expressed in another theory. We establish several negative results which are relevant to preprocessing and normalisation of string constraints in practice. Some of our results have strong connections to important open problems regarding word equations and the theory of string solving.

Weak Bisimulation Finiteness of Pushdown Systems with Deterministic Epsilon-Transitions Is 2-EXPTIME-Complete

Stefan Göller and Paweł Parys

Abstract. We consider the problem of deciding whether a given pushdown system all of whose epsilon-transitions are deterministic is weakly bisimulation finite, that is, whether it is weakly bisimulation equivalent to a finite system. We prove that this problem is 2-EXPTIME-complete. This consists of three elements: First, we prove that the smallest finite system that is weakly bisimulation equivalent to a fixed pushdown system, if exists, has size at most doubly exponential in the description size of the pushdown system. Second, we propose a fast algorithm deciding whether a given pushdown system is weakly bisimulation equivalent to a finite system of a given size. Third, we prove 2-EXPTIME-hardness of the problem. The problem was known to be decidable, but the previous algorithm had Ackermannian complexity (6-EXPSPACE in the easier case of pushdown systems without epsilon-transitions); concerning lower bounds, only EXPTIME-hardness was known.

What Can Oracles Teach us About the Ultimate Fate of Life?

Ville Salo and Ilkka Törmä

Abstract. We settle two long-standing open problems about Conway's Life, a two-dimensional cellular automaton. We solve the Generalized grandfather problem: for all $n \geq 0$, there exists a configuration that has an nth predecessor but not an (n + 1)st one. We also solve (one interpretation of) the Unique father problem: there exists a finite stable configuration that contains a finite subpattern that has no predecessor patterns except itself. In particular this gives the first example of an unsynthesizable still life. The new key concept is that of a spatiotemporally periodic configuration (agar) that has a unique chain of preimages; we show that this property is semidecidable, and find examples of such agars using a SAT solver. Our results about the topological dynamics of Game of Life are as follows: it never reaches its limit set; its dynamics on its limit set is chain-wandering, in particular it is not topologically transitive and does not have dense periodic points; and the spatial dynamics of its limit set is non-sofic, and does not admit a sublinear gluing radius in the cardinal directions (in particular it is not block-gluing). Our computability results are that Game of Life's reachability problem, as well as the language of its limit set, are PSPACE-hard.

Universal Complexity Bounds Based on Value Iteration and Application to Entropy Games

Xavier Allamigeon, Stephane Gaubert, Ricardo Katz, and Mateusz Skomra

Abstract. We develop value iteration-based algorithms to solve in a unified manner different classes of combinatorial zero-sum games with mean-payoff type rewards. These algorithms rely on an oracle, evaluating the dynamic programming operator up to a given precision. We show that the number of calls to the oracle needed to determine exact optimal (positional) strategies is, up to a factor polynomial in the dimension, of order R/sep, where the "separation" sep is defined as the minimal difference between distinct values arising from strategies, and R is a metric estimate, involving the norm of approximate sub and super-eigenvectors of the dynamic programming operator. We illustrate this method by two applications. The first one is a new proof, leading to improved complexity estimates, of a theorem of Boros, Elbassioni, Gurvich and Makino, showing that turn-based mean-payoff games with a fixed number of random positions can be solved in pseudo-polynomial time. The second one concerns entropy games, a model introduced by Asarin, Cervelle, Degorre, Dima, Horn and Kozyakin. The rank of an entropy game is defined as the maximal rank among all the ambiguity matrices determined by strategies of the two players. We show that entropy games with a fixed rank, in their original formulation, can be solved in polynomial time, and that an extension of entropy games incorporating weights can be solved in pseudo-polynomial time under the same fixed rank condition.

Regular Path Queries in MillenniumDB

Domagoj Vrgoc and Carlos Rojas

Abstract. MillenniumDB is a recently published, open-source, graph database engine based on traditional relational storage mechanisms, and state-of-the art query execution techniques such as worst-case optimal join algorithms and path search guided by automata. Here we will concentrate on MillenniumDB's algorithms for executing regular path queries, and the extensions of classical graph reachability algorithms it uses to tackle this problem.

On the Skolem Problem for Reversible Sequences

George Kenison

Abstract. Given an integer linear recurrence sequence $<X_n>$, the Skolem Problem asks to determine whether there is a natural number n such that $X_n = 0$. Recent work by Lipton, Luca, Nieuwveld, Ouaknine, Purser, and Worrell proved that the Skolem Problem for a class of reversible sequences is decidable up to order seven. Here we give an alternative proof of their result. Our novel approach employs a powerful result for Galois conjugates that lie on two concentric circles due to Dubickas and Smyth.

Contents

Invited Papers

SAT-Based Invariant Inference and Its Relation to Concept Learning

Yotam M. Y. Feldman and Sharon Shoham[✉]

Tel Aviv University, Tel Aviv, Israel
yotam.feldman@gmail.com, sharon.shoham@gmail.com

Abstract. This paper surveys results that establish formal connections and distinctions between SAT-based invariant inference and exact concept learning with queries, showing that learning techniques and algorithms can clarify foundational questions, illuminate existing algorithms, and suggest new directions for efficient invariant inference.

1 Introduction

SAT-based invariant inference algorithms such as IC3/PDR [4,8] and Interpolation [25] have proven to be extremely successful in practice and have attracted tremendous interest in recent years. However, the essence of their practical success and their performance guarantees are far less understood. In a series of papers [10–13] we set out to investigate these topics and provide new insights into the principles and complexity of SAT-based invariant inference. This paper surveys one of the key avenues pursued in these works, which focuses on the similarities and discrepancies between SAT-based invariant inference and exact concept learning from queries for propositional formulas, both as a way to explain and analyze existing inference algorithms, and as way to develop new algorithms.

Exact learning with queries [2] is one of the fundamental fields of theoretical machine learning. There, a *learner* (an algorithm) needs to learn an unknown concept, e.g., a formula from some class of formulas, with the help of a *teacher* who can answer certain queries about the concept. Typical queries include membership queries: "is a certain example a member of the desired concept?", and equivalence queries: "is a certain candidate the desired concept?". The theory of exact concept learning is well developed and provides ample efficient algorithms for learning different classes of concepts with different kinds of queries.

The goal of SAT-based invariant inference is also to learn a formula—an inductive invariant. Further, the way an inference algorithm uses a SAT solver to check inductiveness and bounded reachability w.r.t. the transition relation in the process of constructing candidate inductive invariants bears strong resemblance to how a learning algorithm uses the teacher to check equivalence or membership in concept learning.

The first step towards understanding invariant inference from the perspective of learning is to distill this connection and study it in a rigorous way that enables

A. W. Lin et al. (Eds.): RP 2022, LNCS 13608, pp. 3–27, 2022.
https://doi.org/10.1007/978-3-031-19135-0_1

a transfer of ideas between the fields. To this end, we introduce a model of *invariant inference with queries* [10]. In this model, the transition relation of the system is only known to an oracle (implemented by a SAT solver), and an inference algorithm can only "query" it by posing queries to the oracle. Through a sequence of queries, the algorithm should gain enough information about the transition relation to be able to find an appropriate invariant. We consider queries that are common in existing invariant inference algorithms: *inductiveness queries*, where the solver is given a candidate invariant α and checks if it is inductive, and their generalization into *Hoare queries*, where the solver is given a precondition formula α, a postcondition formula β and a bound k, and checks if (some state in) β is reachable from (some state in) α in at most k steps (inductiveness queries correspond to Hoare queries with $\alpha = \beta$ and $k = 1$). Hoare queries naturally capture how many invariant inference algorithms use a SAT solver, including major versions of PDR and Interpolation, and so results about the Hoare-query model apply to these algorithms.

A previous learning-based model for invariant inference, ICE learning [15], corresponds to algorithms that use inductiveness queries only. In practice, many algorithms (that historically precede the ICE learning model) use more general Hoare queries to facilitate an incremental construction of invariants in complex syntactic forms. For example, PDR [4,8] incrementally learns clauses in different frames via relative inductiveness checks, and Interpolation learns at each iteration a term of the invariant from an interpolant [25]. We show that this is in fact a significant difference: the Hoare-query model is strictly stronger than inference based solely on presenting whole candidate inductive invariants as in the inductiveness-query model. To this end, in [10], we identify a class of systems where a Hoare query algorithm, which is essentially a simplified version of PDR (and a dual version of Interpolation), can efficiently infer invariants, whereas every inference algorithm in the inductiveness query model requires an exponential number of queries in the worst case. This confirms the intuition from [39] that PDR cannot be implemented within the ICE model.

Having laid the foundations, we set out to compare invariant inference with queries to exact concept learning. We prove that neither membership queries nor equivalence queries to an unknown invariant can be implemented by Hoare queries in general [10]. In particular, even though inductiveness queries can determine if a formula is an inductive invariant, they are still unable to simulate equivalence queries since they can only return a *counterexample to induction*—a *pair* of states such that *if* the first state is part of the invariant *then* so should be the second. The non-implementability result implies that neither inductiveness nor Hoare queries are sufficient for identifying a (positive or negative) example that definitively differentiates the formula from an inductive invariant. This provides a formal justification to the introduction of *implication examples* in the ICE model [15] for learning from examples, as an addition to positive and negative examples.

The inability to implement exact learning queries is unfortunate, as it prevents porting the rich literature of exact learning algorithms and theory to invariant inference. However, we identify a condition, called the *fence condition* [11], that rectifies the situation and makes it possible to simulate certain kinds of exact learning queries in the Hoare-query model. The fence condition requires that the states in the *boundary* of the invariant—states outside of the invariant with a Hamming distance of 1 from states inside the invariant—can reach a bad state in a bounded number of steps. We show in [11] that when the membership and equivalence queries performed by an exact learning algorithm satisfy certain restrictions, it is possible to translate the learning algorithm into an invariant inference algorithm in the Hoare-query model that is always sound (i.e., never returns an incorrect inductive invariant), and enjoys the same complexity if the fence condition holds.

These translations are not just theoretical. In [11] we show that a model-based version due to Chockler et al. and Bjørner et al. [3,7] of McMillan's Interpolation algorithm [25] can be obtained by such a translation from an exact learning algorithm for learning DNF formulas [1], which is efficient for monotone formulas [1,2]. Not only is it fascinating that an inference algorithm turns out to be an incarnation of an earlier algorithm from a different discipline, but the translation also gives rise to a new efficiency result for the interpolation-based algorithm for monotone invariants when the fence condition holds. To the best of our knowledge this is the first result of its kind. The translation is also applicable to an exact learning algorithm for almost-monotone invariants (and its complexity) [5], which leads to the introduction of a new invariant inference algorithm with provable polynomial complexity guarantees for almost-monotone invariants when the fence condition holds.

The aforementioned simulation of exact learning algorithms is only possible when the queries are restricted in a certain way. Some algorithms, such as Bshouty's algorithm [5] for learning CDNF formulas—formulas that have a short CNF representation as well as a short DNF representation—do not meet these restrictions. Nonetheless, we present in [11] another translation that is applicable to *any* exact learning algorithm from membership and equivalence queries, and maintains the complexity of the algorithm if a stronger, *two-sided* fence condition holds.

The question whether it is possible to simulate Bshouty's CDNF algorithm under a (one-sided) fence condition remains open. However, inspired by the CDNF algorithm, and utilizing insights about properties of the boundary of an invariant, in [13], we develop a novel invariant inference algorithm from Hoare queries that is efficient for CDNF invariants under the assumption of a (one-sided) fence condition. Interestingly, this algorithm cannot be viewed as a concept learning algorithm, hinting that invariant inference can not only benefit from exact learning, but can also exceed it.

Not included in this survey is our investigation of PDR in [12] using the monotone theory [5] developed for concept learning. This work does not investigate PDR as a concept learning algorithm, but relates it to key principles used in learning monotone and almost-monotone invariants.

Outline. The rest of the paper is organized as follows. After a brief background in Sect. 2, we define the model of invariant inference from queries and show an exponential gap between inductiveness queries and Hoare queries in Sect. 3 (based on [10]). We contrast invariant inference with exact concept learning from queries in Sect. 4 (based on [10]), and present cases where the gap can be bridged through the fence condition in Sect. 5 (based on [11,13]). We conclude in Sect. 6.

2   Preliminaries

Transition Systems. We consider transition systems defined using propositional logic. Given a propositional vocabulary Σ, a *state* is a valuation to Σ. We denote by $\mathcal{F}(\Sigma)$ the set of well-formed formulas over Σ. A *transition system* is a triple $TS = (Init, \delta, Bad)$ such that $Init, Bad \in \mathcal{F}(\Sigma)$ define the *initial states* and the *bad states*, respectively, and $\delta \in \mathcal{F}(\Sigma \uplus \Sigma')$ defines the *transition relation*, where $\Sigma' = \{x' \mid x \in \Sigma\}$ is a copy of the vocabulary used to describe the post-state of a transition. Given a $\varphi \in \mathcal{F}(\Sigma)$, we denote by φ' the formula obtained from φ by replacing each variable with its counterpart in Σ'.

Safety and Inductive Invariants. A transition system TS is *safe* if all the states that are reachable from the initial states via steps of δ satisfy $\neg Bad$. An *inductive invariant* for TS is a formula $I \in \mathcal{F}(\Sigma)$ such that (i) $Init \implies I$, (ii) $I \wedge \delta \implies I'$, and (iii) $I \implies \neg Bad$ (where $\implies$ denotes validity of implication). A transition system is safe if and only if it has an inductive invariant. When I is not inductive, a *counterexample to induction (cti)* is a pair of states σ, σ' such that $\sigma, \sigma' \models I \wedge \delta \wedge \neg I'$ (where the valuation to Σ' is taken from σ').

Notation. We use formulas and the sets of states that they represent interchangeably. For a state σ, we denote by $cube(\sigma)$ the conjunction of all literals (variables or their negations) that hold in σ.

3   Invariant Inference with Queries

An investigation of SAT-based invariant inference and its relation to concept learning was initiated in [10], by identifying the common SAT queries carried out by existing algorithms, and introducing corresponding query models. A query-based approach allows to compare different invariant inference algorithms both to each other and to concept-learning algorithms that use queries.

In this section we define the invariant inference problem, the basic notions of queries and query-based inference algorithms, and the query models considered in this survey: Inductiveness and Hoare, which capture existing SAT-based invariant inference algorithms.

Invariant inference can be formulated as follows.

Definition 1 (Inductive Invariant Inference from Class of Invariants).
For a class of transition systems $\mathcal{P}$ and a class of invariants $\mathcal{L}$, inductive invariant inference is the problem: Given a transition system $TS \in \mathcal{P}$ over Σ, find an inductive invariant $I \in \mathcal{L}$ for TS or determine that none exists.

When $\mathcal{L}$ is omitted, we mean that, for every Σ, it includes all formulas in $\mathcal{F}(\Sigma)$.

3.1   Inference with Queries

In the setting of invariant inference with queries, an algorithm accesses the transition relation through queries—corresponding to SAT queries performed by existing algorithms—but cannot read the transition relation directly. This *black-box* model reflects the way typical SAT-based invariant inference algorithms use the transition relation only in their SAT queries, as opposed to *white-box* algorithms that analyze the code directly. A black-box model of inference algorithms facilitates an analysis of the *information* of the transition relation the algorithm acquires. The advantage is that such an information-based analysis sidesteps open computational complexity questions, and therefore results in unconditional lower bounds on the complexity of SAT-based algorithms captured by the model.

Queries of the transition relation are modeled in the following way. A *query oracle Q* is an oracle that accepts a transition relation δ, as well as additional inputs, and returns some output. The additional inputs and the output, together also called the *interface* of the oracle, depend on the query oracle under consideration. A *family* of query oracles is a set of query oracles with the same interface.

Definition 2 (Inference algorithm in the query model). *An* inference algorithm from queries, *denoted* $\mathcal{A}^Q(Init, Bad, [\delta])$, *is an algorithm defined w.r.t. a query oracle Q that solves the invariant inference problem for* $(Init, \delta, Bad)$, *given:*

- *access to the query oracle Q,*
- *the set of initial states (Init) and bad states (Bad);*
- *the transition relation δ, encapsulated—hence the notation $[\delta]$—meaning that the algorithm cannot access δ (not even read it) except for extracting its vocabulary; δ can only be passed as an argument to the query oracle Q.*

In the sequel, we consider two different families of query oracles: inductiveness and Hoare, representing different ways of obtaining information about the transition relation.

Time and Query Complexity. Much like a SAT solver, the query oracles solve NP-complete problems. When analyzing the complexity we consider each query as a single step, and count the number of queries and also the time of other steps the algorithm performs. (In lower bounds, we often only report on the query complexity, which in itself provides a lower bound on the time complexity.) We analyze the complexity in a *worst-case* model w.r.t. the possible transition systems in the class of interest as well as w.r.t. the possible query oracles in the family (the worst-case analysis is motivated by the property that in SAT-based algorithms, the oracle is implemented by a SAT solver, which the algorithm does not control). For a class of transition systems $\mathcal{P}$, the (time or query) complexity of $\mathcal{A}$ w.r.t. a query oracle family $\mathcal{Q}$ is defined as

$$\sup_{Q \in \mathcal{Q}} \sup_{\substack{(Init,\delta,Bad) \in \mathcal{P}, \\ |\Sigma|=n}} \Phi(\mathcal{A}^Q(Init, [\delta], Bad))$$

where $\Phi(\mathcal{A}^Q(Init, [\delta], Bad))$ measures the complexity (either number of steps or number of queries) of $\mathcal{A}$ given oracle $Q \in \mathcal{Q}$ and input $(Init, \delta, Bad) \in \mathcal{P}$. (These numbers might be infinite.)

3.2   The Inductiveness-Query Model

The first query model we consider only allows an algorithm to check inductiveness of a candidate invariant:

Definition 3 (Inductiveness-Query Model).   *An* inductiveness-query ora-*cle is a query oracle* $\mathcal{I}$ *such that for every* δ *and* $\alpha \in \mathcal{F}(\Sigma)$ *satisfying* $Init \Longrightarrow \alpha$ *and* $\alpha \Longrightarrow \neg Bad$,

- $\mathcal{I}(\delta, \alpha) = true$ *if* $\alpha \wedge \delta \Longrightarrow \alpha'$, *and*
- $\mathcal{I}(\delta, \alpha) = (\sigma, \sigma')$ *such that* $(\sigma, \sigma') \models \alpha \wedge \delta \wedge \neg\alpha'$ *otherwise.*

An algorithm in the inductiveness-query model, *also called an* inductiveness-query algorithm, *is an inference from queries algorithm expecting any inductiveness query oracle.*

Inductiveness-query oracles form a family *of oracles since different oracles can choose different* (σ, σ') *for each* δ, α.

ICE Learning and Inductiveness-Queries. The inductiveness-query model is closely related to ICE learning [15], except here the learner is provided with full information on *Init, Bad* instead of positive and negative examples (and the algorithm refrains from querying on candidates that do not include *Init* or do not exclude *Bad*). This model captures several interesting algorithms, including include Houdini [14] and symbolic abstraction [30, 35], as well as designated algorithms [15, 16]. Our complexity definition in the inductiveness-query model being the worst-case among all possible oracle responses is in line with the analysis of strong convergence in Garg et al. [15]. Hence, lower bounds on the query complexity in the inductiveness query model imply lower bounds for the strong convergence of ICE learning.

3.3   The Hoare-Query Model

The Hoare-query model captures SAT-based invariant inference algorithms querying the reachability of one set of states from a possibly different set of states through a sequence of at most k-steps of the transition relation, for a fixed k.

Definition 4 (Hoare-Query Model). *A* Hoare-query oracle *is a query oracle $\mathcal{H}$ such that for every $\delta, \alpha, \beta \in \mathcal{F}(\Sigma)$, and k,*

- $\mathcal{H}^{(k)}(\delta, \alpha, \beta) = true \ if \ \alpha(\Sigma^0) \wedge \delta(\Sigma^0, \Sigma^1) \wedge \ldots \wedge \delta(\Sigma^{k-1}, \Sigma^k) \implies \bigwedge_{i=1}^{k} \beta(\Sigma^i),$
 where $\Sigma^0, \ldots, \Sigma^k$ are $k+1$ distinct copies of the vocabulary, and
- $\mathcal{H}^{(k)}(\delta, \alpha, \beta) = (\sigma_0, \ldots, \sigma_k)$ *such that* $\sigma_0, \ldots, \sigma_k \models \alpha(\Sigma^0) \wedge \delta(\Sigma^0, \Sigma^1) \wedge \ldots \wedge \delta(\Sigma^{k-1}, \Sigma^k) \wedge \bigvee_{i=1}^{k} \neg\beta(\Sigma^k),$ *otherwise.*

An algorithm in the Hoare-query model, *also called a* Hoare-query algorithm, *is an inference from queries algorithm expecting any Hoare-query oracle, where k is bounded by a polynomial in n in all queries.*

Hoare-query oracles form a *family* of oracles since different oracles can choose different counterexample traces $(\sigma_0, \ldots, \sigma_k)$ for every δ, α, β, k.

Example: PDR as a Hoare-Query Algorithm. The Hoare-query model captures the prominent PDR algorithm, facilitating its theoretical analysis. In general, PDR maintains a sequence of frames $F_0, F_1, \ldots$ such that $F_0 = Init$, $F_i \implies F_{i+1}$, $F_i \wedge \delta \implies F'_{i+1}$ and $F_i \implies \neg Bad$ (for every i). These properties ensure that if at some point $F_{i+1} \implies F_i$ then F_i is an inductive invariant. To update the frames, PDR accesses the transition relation via checks of unreachability in one step and counterexamples to those checks. These operations are captured in the Hoare query model by checking $\mathcal{H}^{(1)}(\delta, F, \alpha)$ or $\mathcal{H}^{(1)}(\delta, F \wedge \alpha, \alpha)$. This is illustrated using Algorithm 1 which roughly corresponds to PDR with just one frame. The only accesses to δ are in lines 5, 6 and 9, which are all done through the Hoare-query oracle, showing that Algorithm 1 is a Hoare-query algorithm. The (basic) full PDR can similarly be modeled as a Hoare-query algorithm [10]. Furthermore, the Hoare-query model is general enough to express a broad range of PDR variants that differ in the way they use such checks but still access the transition relation only through such queries.[1]

Example: Interpolation-Based Inference as a Hoare-Query Algorithm. Another operation supported by SAT solvers is *interpolation*. Interpolation has been introduced to invariant inference by McMillan [25], and extended in many works since [18, 22, 26, 37, 38]. Interpolation algorithms infer invariants from facts obtained from bounded unreachability of the bad states, checked by Hoare queries of the form $\mathcal{H}^{(k)}(\delta, F, \neg Bad)$. In McMillan's original paper these facts are *interpolants* extracted from a resolution proof computed by the solver. As such, to account for McMillan's original interpolation-based inference algorithm [25], the oracle also needs to return an interpolant when the Hoare query checking unreachability of *Bad* returns *true*. This model was investigated in [10]. Our focus here is on model-based interpolation [3, 7], as displayed in Algorithm 2, for which such an extension is not necessary—model-based interpolation computes

---

[1] A notable exception is ternary simulation [8], which is not a SAT-based operation. However, the query model can be extended to support it while maintaining our results.

interpolants as part of the inference procedure (from bounded unreachability and counterexamples), rather than inside the solver (from the proof of bounded unreachability itself). Algorithm 2 is a Hoare-query algorithm: the only accesses to δ are in lines 3, 4 and 9, and all invoke the Hoare-query oracle.

3.4   Hoare-Queries vs. Inductiveness-Queries

Inductiveness queries are specific instances of Hoare queries, where the precondition and postcondition are the same, and reachability is examined along a single step of the transition relation ($k = 1$). Therefore, inductiveness-query algorithms can be simulated by Hoare-query algorithms. This raises the question whether the seemingly more general Hoare queries are indeed so. In this section we answer this question affirmatively and show that the Hoare query model (Definition 4) is *strictly stronger* than the inductiveness query model (Definition 3), even when $k = 1$. To this end we show that there exists a class of transition systems for which a simple Hoare-query algorithm can infer invariants in polynomial time, but every inductiveness-query algorithm requires an exponential number of queries.

The exponential gap between the Hoare-query model and the inductiveness query model is summarized by the following theorem:

Theorem 1 ([10]). *There exists a class of transition systems $\mathcal{M}_E$ for which*

- *invariant inference has* polynomial *time complexity in the Hoare-query model, but*
- *every inference algorithm in the inductiveness-query model requires an* exponential *query complexity.*

The class $\mathcal{M}_E$ consists of *maximal systems* for *monotone* CNF invariants together with certain unsafe systems.[2] We refer the reader to [10] for the precise definition of $\mathcal{M}_E$ and to the proof of the lower bound in the inductiveness-query model. Here we only highlight two properties of the safe systems in $\mathcal{M}_E$ that facilitate efficient inference in the Hoare-query model: *maximality* of the system, defined below, and existence of a polynomial *monotone* invariant. Beyond establishing the upper bound in the Hoare-query model, these properties also spur the research on efficient inference discussed in subsequent sections.

Definition 5 (Monotone Invariants). *We denote by Mon-CNF the class of CNF formulas where variables appear only positively and where, for a vocabulary Σ with $n = |\Sigma|$, the number of clauses in formulas over Σ is bounded by $p(n)$, for a fixed polynomial $p(\cdot)$.*

Definition 6 (Maximal System). *Let Init, Bad $\not\equiv$ false and let φ be a formula such that Init $\implies \varphi$ and $\varphi \implies \neg$Bad. The maximal transition system for φ is $(Init, \delta_\varphi^{\mathcal{M}}, Bad)$ where $\delta_\varphi^{\mathcal{M}} = \varphi \to \varphi'$.*

---

[2] In [10], the invariants are *antimonotone* rather than monotone; the algorithm establishing the upper bound is efficient also for monotone invariants, and the proof of the lower bound can also be adapted to monotone invariants.

A maximal transition system is illustrated as follows:

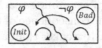

Note that $\delta_\varphi^{\mathcal{M}}$ goes from any state satisfying φ to any state satisfying φ, and from any state satisfying $\neg\varphi$ to all states, good or bad. $\delta_\varphi^{\mathcal{M}}$ is *maximal* in the sense that it allows all transitions that do not violate φ being an inductive invariant.

In $\mathcal{M}_E$ we consider maximal systems for every formula in Mon-CNF, together with an unsafe transition system whose transition relation is *true* for each vocabulary. In particular, this means that every safe transition system in $\mathcal{M}_E$ has an inductive invariant in Mon-CNF (and others have no inductive invariant). Therefore, solving invariant inference for $\mathcal{M}_E$ without restricting the class of invariants coincides with restricting it to $\mathcal{L} = $ Mon-CNF; this is important in Sect. 4.1 when comparing the complexity of invariant inference to exact concept learning.

Upper Bound for Hoare-Query Algorithms for Maximal Systems w.r.t. Monotone Invariants. The upper bound is obtained by a simple algorithm, called PDR-1, that can find inductive invariants for safe systems in $\mathcal{M}_E$ with a polynomial number of Hoare queries. PDR-1, depicted in Algorithm 1, is a backward-reachability algorithm, operating by repeatedly checking for the existence of a counterexample to induction, and obtaining one when it exists. The invariant is then strengthened by conjoining the candidate invariant with the negation of a subset of the cube of the pre-state: starting with $cube(\sigma)$, which is a conjunction of literals that holds only on σ, a subset of the literals leaves a smaller conjunction, which represents a larger set of states, thereby "generalizing" σ. Generalization is performed by dropping a literal from the cube whenever the remaining conjunction does not hold for any state reachable in at most one step from *Init*. The result is a minimal conjunction whose negation does not exclude any state reachable in at most one step. This might exclude reachable states in general transition systems, but not in maximal systems, since maximality ensures that their diameter is one.

Algorithm 1. PDR-1 in the Hoare-query model

```
1:  procedure PDR-1(Init, [δ], Bad)
2:      I ← ¬Bad
3:      if H^(1)(δ, Init, ¬Bad) ≠ true then
4:          unsafe
5:      while H^(1)(δ, I, I) ≠ true do                    // I not inductive
6:          (σ, σ') ← H^(1)(δ, I, I)           // counterexample to induction of I
7:          d ← cube(σ)
8:          for l ∈ cube(σ) do
9:              t ← d \ {l}
10:             if Init ⟹ ¬t and H^(1)(δ, Init, ¬t) then    // Init ⟹ ¬t ∧ Init ∧ δ ⟹ ¬t'
11:                 d ← t
12:         I ← I ∧ ¬d
13:     return I
```

Maximality therefore allows to determine if a state needs to be part of the invariant by a simple Hoare-query, examining reachability in 1-step, and ensures that generalization returns a prime consequence of the invariant (a clause implied by the invariant which is not strictly weaker than any other clause implied by the invariant). Efficiency of the algorithm results from the monotonicity of the CNF invariants, which lets PDR-1 efficiently reconstruct them as the conjunction of their prime consequences, via a theorem that goes back to Quine [29].

However, our lower bound for the inductinvess-query model for the same class of transition systems and invariants shows that this incremental process inherently relies on rich Hoare queries.

3.5 PDR and Interpolation-Based Inference Cannot Be Implemented with Inductiveness Queries

PDR-1, the Hoare-query algorithm we use to establish the exponential gap, is essentially PDR with a single frame[3]. Hence, building on the proof of Theorm 1, which shows that no inductiveness-query algorithm can simulate PDR-1 on the class $\mathcal{M}_E$, we conclude that PDR cannot be efficiently simulated in the inductiveness-query model:

Theorem 2. *There is no inductiveness-query algorithm that solves invariant inference with a number of inductiveness queries that has at most polynomial overhead on the number of Hoare queries performed by PDR.*

A similar result applies to interpolation-based inference: the exponential gap between the inductiveness and Hoare query models can also be established for maximal systems for Mon-DNF invariants (defined similarly to Mon-CNF, as the class of DNF formulas where variables appear only positively and where the number of terms is bounded by $p(n)$), in which case the upper bound is obtained by an algorithm dual to PDR-1, the model-based interpolation-based algorithm displayed in Algorithm 2 with a reachability bound of $k = 1$. This shows that Hoare-queries are inherent to both PDR and interpolation-based inference in the sense that neither can be implemented with inductiveness queries only, confirming the intuition from [39] regarding PDR. (Profound differences between PDR and interpolation manifest when PDR uses more than one frame and interpolation uses $k > 1$, a topic we explored in [12].)

4 Invariant Learning and Concept Learning with Queries

Query-based models of invariant inference highlight its similarity to exact concept learning with queries. What are the connections and differences between

---

[3] To be precise, in PDR, counterexamples are states that reach a bad state, whereas PDR-1 uses counterexamples to induction, but these coincide in maximal systems; additionally, PDR may use an additional frame to discover the counterexamples and one more to detect convergence.

concept-learning *formulas* in $\mathcal{L}$ and learning *invariants* in $\mathcal{L}$? Can concept learning algorithms be translated to inference algorithms? These questions have spurred much research [9,15,16,20,21,23,27,31–34]. In this section we study these questions from the perspective of our aforementioned results.

In *exact concept learning* [2], an algorithm's task is to identify an unknown formula[4] ψ using queries it poses to a *teacher*. The most studied queries are:

– *Membership*: The algorithm *chooses* a state σ, and the teacher answers whether $\sigma \models \psi$; and
– *Equivalence*: The algorithm *chooses* a candidate θ, and the teacher returns true if $\theta \equiv \psi$ or a differentiating counterexample otherwise: a σ s.t. $\sigma \not\models \theta, \sigma \models \psi$ or $\sigma \models \theta, \sigma \not\models \psi$.

In this section, we compare invariant inference to exact concept learning and show: (1) that classical queries in exact concept learning cannot be efficiently implemented as queries in order to find an unknown inductive invariant, and (2) that ICE-learning is provably harder than classical learning: namely, that, as advocated by Garg et al. [15], learning from counterexamples to induction is inherently harder than learning from examples labeled positive or negative.

4.1   Complexity Comparison

This section compares the complexity of inferring formulas to concept-learning the same class of formulas.

Theorem 1 effectively studies the complexity of inferring $\mathcal{L} = $ Mon-CNF invariants using Hoare/inductiveness queries for *maximal systems*. The next theorem studies the complexity for *general systems*:

Theorem 3. *Every Hoare-query inference algorithm solving invariant inference for the class of all propositional transition systems and the class of invariants* $\mathcal{L} = $ *Mon-CNF has query complexity of* $2^{\Omega(n)}$*, where* $n = |\Sigma|$*.*

We emphasize that the lower bound considers inference of short, polynomial, invariants, which ensures that the exponential complexity is not an artifact of the length of the invariant, but, rather, of the need to infer it. We also point out that in the more standard setting, when the algorithm is not restricted to access the transition relation only through Hoare queries, the computational complexity of inferring invariants of polynomial length is Σ_2^P-complete (NP-complete with access to a SAT solver as oracle), as shown in [10] (strengthening a similar hardness result by Lahiri and Qadeer for inferring invariants over a template [24]).

Table 1 displays these results for invariant inference with a query oracle, and compares them with known complexity results for exact concept learning. For the sake of the comparison, the table maps inductiveness queries to equivalence

---

[4] In general, a concept is a set of elements; here we focus on logical concepts.

Table 1. Concept vs. invariant learning: complexity of learning Mon-CNF

Invariant inference			Concept learning	
	Maximal systems	General systems		
Inductiveness	Exponential (Theorem 1)	Exponential (Theorem 3)	Equivalence	Subexponential[1] / Polynomial[2] [2,17]
Hoare	Polynomial (Theorem 1)	Exponential (Theorem 3)	Equivalence + Membership	Polynomial [2]

[1] Proper learning
[2] With exponentially long candidates

queries (as these are similar at first sight) and maps the more powerful setting of Hoare queries to the more powerful setting of equivalence together with membership queries.

The comparison in the table demonstrates that *invariant inference in general systems is harder* than exact learning. The implications of the complexity gaps are elaborated in Sect. 4.2. The complexity gap is eliminated when considering only *maximal systems*, which is the source of the upper bound in Theorem 1. However, that is true only for the Hoare-query model, and gaps remain when considering only inductiveness queries; this is elaborated in Sect. 4.3.

4.2   Invariant Learning Cannot Be Reduced to Concept Learning

This section builds on the above complexity comparison to check which concept learning queries can be simulated and used in invariant inference.

Table 2. Concept vs. invariant learning: implementability of concept learning queries

	Maximal systems		General systems	
	Inductiveness	Hoare	Inductiveness	Hoare
Equivalence	✗	✓	✗	✗
Membership	✗	✓	✗	✗

Table 2 summarizes our results for the possibility and impossibility of simulating concept learning algorithms in invariant learning with queries. This table depicts implementability (✓) or unimplementability (✗) of membership and equivalence queries used in concept learning through inductiveness and Hoare queries used in learning invariants for maximal systems and for general systems.

Formally, the implementability of a (concept learning) query in a class of transition systems $\mathcal{P}$ means that for every class of invariants $\mathcal{L}$ there is an inference algorithm that, given a transition system $TS \in \mathcal{P}$ that admits some (unknown) invariant $I \in \mathcal{L}$, correctly answers the query w.r.t. I with a polynomial number of queries in the respective model.

The proofs of impossibilities are based on the differences in complexity from Table 1 for $\mathcal{L} = $ Mon-CNF. The only possibility result in the table is of simulating equivalence and membership queries using Hoare queries over maximal systems (for every $\mathcal{L}$); the idea is that a Hoare query $\mathcal{H}^{(1)}(\delta_\varphi^\mathcal{M}, Init, \neg cube(\sigma)) \overset{?}{\neq} true$ implements a membership query on σ, thanks to fact that the inductive invariant is exactly the set of states reachable in one step. Such a membership query (together with an inductiveness query) can also be used to implement an equivalence query, specifically to convert a counterexample to induction into a differentiating counterexample as required when answering an equivalence query negatively: given the counterexample to induction $(\sigma, \sigma') = \mathcal{I}(\delta_\varphi^\mathcal{M}, \theta)$, use a membership query to determine if $\sigma \not\models I$ or $\sigma' \models I$, and return σ or σ' accordingly. We pick up on these ideas for implementing membership queries in Sect. 5, with more sophisticated translations that are related to more realistic algorithms.

4.3   Counterexamples in Invariant Learning Are Inherently Ambiguous

As we have seen, equivalence queries cannot be implemented using inductiveness queries, even in the simple case of maximal systems. The reason is that when the query fails—returns "not inductive" or "not equivalent"—then the counterexample provided to the inference algorithm is inherently weaker than the counterexample for the learning algorithm. In inference, the result is a counterexample to induction (an implication example, in the terminology of Garg et al. [15]), which is a *pair* of examples (σ, σ'), where σ is a negative example *or* σ' is a positive example, but there is no indication in the query itself of which is the case. In contrast, in classical equivalence queries, the counterexample is a single state σ, and it is in effect labelled—by checking whether the proposed candidate is satisfied by σ or not the learner can tell whether σ is a positive or negative example.

This discrepancy can be reformulated in the context of concept learning, as the difference between classical learning from equivalence queries (using labeled examples) and ICE learning [15], in which (essentially) the result of an equivalence query is an implication example. We have thus obtained a complexity result separating the two:

Corollary 1. *There exists a class of formulas $\mathcal{L}$ that can be learned using a subexponential number of equivalence queries, but requires an exponential number of ICE-equivalence queries.*

This result quantitatively corroborates the difference between *counterexamples to induction* and *examples labeled positive or negative*, a distinction advocated by Garg et al. [15].

5   From Exact Learning to Invariant Inference via the Fence Condition

We have seen that Hoare-queries cannot, in general, simulate equivalence and membership queries used in exact concept learning, but can do so for maximal systems. This is somewhat disheartening; it would have been much nicer to

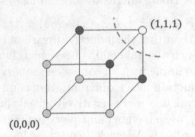

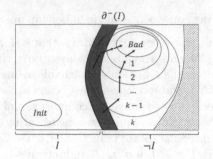

Fig. 1. The (outer) boundary of an invariant $I = x \wedge y \wedge z$, denoting the singleton set containing the far-top-right vertex of the 3-dimensional Boolean hypercube, $\{(1,1,1)\}$. Its neighbors are I's boundary (depicted in red): $\{(1,1,0),(1,0,1),(0,1,1)\}$. The rest of the vertices are in $\neg I$ but not in the boundary (depicted in gray). (Illustration inspired by Fig. 2.1 [28].)

Fig. 2. An illustration of the fence condition. The boundary $\partial^-(I)$ of the invariant (the states in $\neg I$ nearest to I, in red) are backwards k-reachable (reach a bad state in k steps, for example by the transitions depicted by the arrows), but not all states in $\neg I$ are backwards k-reachable (or even backwards reachable at all, in the dotted area).

apply an algorithm for learning a class of formulas $\mathcal{L}$ to the problem of inferring invariants from $\mathcal{L}$. In this section we present the *fence condition*, which was introduced in [11]. This condition relaxes the maximality property, and we show that it facilitates simulation of certain exact concept learning algorithms. In particular, we obtain the model-based interpolation-based algorithm of [3,7]—as well as a new algorithm extending it—by a translation from exact learning algorithms that satisfy certain restrictions. The translation also lets us import complexity upper bounds for the obtained inference algorithms from the learning algorithms, revealing new results on the efficiency of inference algorithms provided that the fence condition holds. We further show that when a two-sided fence condition holds, *every* algorithm for exact learning from equivalence and membership queries can be transformed to a Hoare-query inference algorithm.

5.1   The Boundary of Inductive Invariants

The fence condition relates reachability in the transition system and the geometric notion of the *boundary* of the invariant.

Definition 7 (Boundary). *Let I be a set of states. Then the (outer) boundary of I, denoted $\partial^-(I)$, is the set of states $\sigma^- \not\models I$ s.t. there is a state σ^+ that differs from σ^- in exactly one variable, and $\sigma^+ \models I$.*

Definition 8 (Backwards k-Fenced). *For a transition system $(Init, \delta, Bad)$, an inductive invariant I is backwards k-fenced for $k \in \mathbb{N}$ if every state in $\partial^-(I)$ can reach Bad in at most k steps.*

More explicitly, an invariant I is backwards k-fenced if every state in $\neg I$ that has a Hamming neighbor in I (these are the states in the outer boundary of I) can reach Bad in at most k steps. For an illustration of the boundary and the fence condition see Figs. 1 and 2.

Example 1. In a maximal system, the (unique) inductive invariant I is backwards 1-fenced, since every state that is not part of I, in particular a state in the outer boundary, has a transition to every other state, including Bad.

In every system, this condition holds for at least one inductive invariant and for some finite k: the weakest inductive invariant, which allows all states but those that can reach Bad in any number of steps, satisfies the condition with the co-diameter, the number of steps that takes for all states that can reach Bad to do so.

Lemma 1. *Every safe transition system $TS = (Init, \delta, Bad)$ admits an inductive invariant $gfp = \{\sigma \mid \forall \sigma' \in Bad. \ (\sigma, \sigma') \notin \delta^*\}$ that is backwards k-fenced for k that is the co-diameter: the minimal k such that for every state σ: $(\exists \sigma' \in Bad. \ (\sigma, \sigma') \in \delta^*) \implies (\exists \sigma'' \in Bad. \ (\sigma, \sigma'') \in \delta^{\leq k})$.*

While this lemma shows the existence of a backwards-fenced invariant through the gfp and co-diameter, the k-fence condition is more liberal: it can hold also for an invariant when not *every* state in $\neg I$ reaches Bad in k steps (or at all), and only the states in $\partial^-(I)$ do. An example demonstrating this follows.

Example 2. Consider an example of a (doubly)-linked list traversal, using i to traverse the list backwards, modeled via predicate abstraction following Itzhaky et al. [19]. The list starts at h. Initially, i points to some location that may or may not be part of the list, and in each step the system goes from i to its predecessor, until that would reach x. We write $\mathbf{s} \rightsquigarrow r$ to denote that r is reachable from $\mathbf{s}$ by following zero or more links. Consider the initial assumption $\mathbf{h} \rightsquigarrow \mathbf{x}$, but $\mathbf{i} \not\rightsquigarrow \mathbf{x}$ (it may be that $\mathbf{x} \rightsquigarrow \mathbf{i}$, or that i is not at all in the list). The bad states are those where $\mathbf{i} = \mathbf{h}$.

An inductive invariant for this system is $\mathbf{h} \rightsquigarrow \mathbf{x} \wedge \neg \mathbf{i} \rightsquigarrow \mathbf{x}$. In predicate abstraction, we may take the predicates $p_{h,x} = \mathbf{h} \rightsquigarrow \mathbf{x}$, $p_{i,x} = \mathbf{i} \rightsquigarrow \mathbf{x}$, and write $I = p_{h,x} \wedge \neg p_{i,x}$, which is a DNF invariant with one term. Hence $\neg I \equiv \neg p_{h,x} \vee p_{i,x}$. The outer boundary $\partial^-(I)$ consists of the states (1) $p_{h,x} = false, p_{i,x} = false$ and (2) $p_{h,x} = true, p_{i,x} = true$. Both states are in fact bad states under the abstraction: both include a state where $\mathbf{i} = \mathbf{h}$, from which x is unreachable (in (1)) or reachable (in (2)). Thus, I is backwards k-fenced for every $k \geq 0$.

In contrast, not all the states in $\neg I$ reach bad states (in particular, I is not the gfp): the state $p_{h,x} = false, p_{i,x} = true$ abstracts only states where $\mathbf{h} \not\rightsquigarrow \mathbf{i}$, and this remains true after going to the predecessor of i. This shows that the fence condition may hold even though I is not the gfp, and not all states in $\neg I$ reach bad states (in k steps or at all).

5.2 Inference from One-Sided Fence and Exact Learning with Restricted Queries

The challenge in harnessing exact learning algorithms for invariant inference is the need to also implement the teacher, which is problematic because the inference algorithm does not know any inductive invariant in advance [15], and, as we have shown in Sect. 4.2, is unable to efficiently implement a classical teacher that answers equivalence and membership queries, even in the more general Hoare-query model. In this section we overcome this problem using the fence condition, provided that the learning algorithm satisfies some conditions.

First, for equivalence queries, as discussed in Sect. 4.3, inductiveness queries can determine if a candidate formula is an inductive invariant. However, when it is not, the difficulty is the ambiguity of counterexamples to induction (σ, σ'), which makes it difficult to know which of σ or σ' should be returned to the learner as an example that differentiates the candidate from the invariant. We circumvent this problem by simply considering algorithms that query only on candidates which are underapproximations of the target I:

Lemma 2 (Implementing positive equivalence queries). *Let* $(Init, \delta,$ *Bad) be a transition system and* I *an inductive invariant. Given* θ *such that* $\theta \implies I$, *it is possible to decide whether* θ *is an inductive invariant or provide a counterexample* $\sigma \models I, \sigma \not\models \theta$, *by*

- *checking whether there is a counterexample* $\sigma' \models Init \wedge \neg\theta$ *and returning* σ' *if one exists; and*
- *using an inductiveness query* $\mathcal{I}(\delta, \theta)$ *to check whether there is a counterexample to induction* $(\sigma, \sigma') \models \theta \wedge \delta \wedge \neg\theta'$, *and returning* σ' *if one exists.*

Otherwise, θ *is an inductive invariant.*

Note that $\theta \not\equiv I$ could be an inductive invariant, which does not amount to an equivalence query *per se*, but then the algorithm has already found an inductive invariant and can stop.

To implement membership queries, we rely on the fence condition. Our main observation here is that if the fence condition holds for I, then it is possible to efficiently implement restricted versions of membership queries:

Lemma 3 (Implementing positive-adjacent membership queries). *Let* $(Init, \delta, Bad)$ *be a transition system and* I *an inductive invariant that is backwards* k-*fenced. Given* σ *s.t.* $\sigma \models I$ *or* $\sigma \in \partial^-(I)$, *it is possible to decide whether* $\sigma \models I$ *by a single Hoare query that checks if* $\mathcal{H}^{(k)}(\delta, cube(\sigma), \neg Bad) = true$ *and answers accordingly.*

In fact, under similar restrictions, we can implement *subset* queries, which generalize membership queries. In a subset query, the learning algorithm *chooses* a formula θ, and the teacher answers whether $\theta \implies I$, where I is the target formula. (A membership query for σ is a subset query with $\theta = \{\sigma\}$.)

Lemma 4 (Implementing positive-adjacent subset queries). *Let $(Init, \delta, Bad)$ be a transition system and I an inductive invariant that is backwards k-fenced. Given θ s.t. $\theta \implies I$ or $\theta \wedge \partial^-(I) \not\equiv \bot$, it is possible to decide whether $\theta \implies I$ by a single Hoare query that checks if $\mathcal{H}^{(k)}(\delta, \theta, \neg Bad) = true$ and answers accordingly.*

A learning algorithm that only performs such queries induces a Hoare-query invariant inference algorithm that simulates it by implementing its queries as above. If the fence condition holds, all queries are answered correctly by the simulation, perhaps except for an equivalence query on θ returning *true* although $\theta \not\equiv I$, but then we have already found an inductive invariant θ and can stop. An additional inductiveness check is used in the inference algorithm before an invariant is returned to ensure that the result is a correct inductive invariant even when the fence condition does not hold. If the latter inductiveness check fails, the algorithm returns "failure".

Corollary 2. *Let C be a class of formulas. Let $\mathcal{A}$ be an exact concept learning algorithm that can identify every $\varphi \in C$ in at most s_1 equivalence queries and s_2 subset queries (including membership queries). Assume further that when $\mathcal{A}$ performs an equivalence query on θ, always $\theta \implies \varphi$, and when $\mathcal{A}$ performs a subset query on θ, always $\theta \implies \varphi$ or $\theta \wedge \partial^-(\varphi) \not\equiv \bot$. Then there exists a Hoare-query invariant inference algorithm that is sound (returns only correct invariants), and, furthermore, can find an inductive invariant for every transition system that admits an inductive invariant $I \in C$ that is backwards k-fenced using at most $s_1 + 1$ inductiveness and s_2 Hoare queries (with argument k), band time the same as of $\mathcal{A}$ up to a constant factor.*

The inference algorithm is sound even when the fence condition does not hold, although in this case successful and efficient convergence is not guaranteed.

Efficient Interpolation-Based Inference of Monotone Invariants Through Exact Learning. Algorithm 2 presents the interpolation-based invariant inference algorithm due to Chockler et al. and Bjørner et al. [3,7], which uses a model-based method for interpolant construction, inspired by IC3/PDR [4,8], rather than constructing interpolants from proofs as in McMillan's original algorithm [25]. Algorithm 2 starts with the candidate invariant $\varphi = Init$, which is gradually increased to include more states. In each iteration, the algorithm performs an inductiveness query ((lines 3 and 4) and terminates if an inductive invariant has been found. If a counterexample to induction (σ, σ') exists, the algorithm generates a term d which includes the post-state σ', and disjoins d to φ to obtain the new candidate (line 11). To obtain d, the algorithm starts with $cube(\sigma')$—the conjunction that exactly captures σ'—and drops literals as long as no state in d can reach a bad state in k steps or less (line 9). These checks are done via Hoare queries. If σ' itself reaches a bad state in k steps, no invariant weaker than φ exists, and the algorithm restarts with a larger bound k (line 6).

Interestingly, Algorithm 2 is essentially the result of the transformation of Corollary 2 applied to the exact learning algorithm for DNF formulas of Aizenstein and Pitt [1] as it appears in Algorithm 3 (similar to the algorithm by Angluin [2]), where EQ denotes an equivalence query and SQ denotes a subset query. The differences are the additional check in line 6 in Algorithm 2, meant to detect failure, and the initialization of φ to *Init* instead of *false*, which can be viewed as an optimization.

Algorithm 2. Interpolation-based inference by term minimization

```
1: procedure MB-ITP(Init, [δ], Bad, k)
2:    φ ← Init
3:    while I(δ, φ) ≠ true do
4:        (σ, σ') ← I(δ, I)
5:        if H^(k)(δ, σ', ¬Bad) ≠ true then
6:            restart with larger k
7:        d ← cube(σ')
8:        for ℓ in d do
9:            if H^(k)(δ, d \ {ℓ}, ¬Bad) = true then
10:               d ← d \ {ℓ}
11:       φ ← φ ∨ d
12:   return I
```

Algorithm 3. Exact concept learning of DNF formulas [1,2,36]

```
1: procedure LEARN-DNF
2:    φ ← false
3:    while EQ(φ) is not ⊥ do
4:        σ' ← EQ(φ)
5:
6:
7:        d ← cube(σ')
8:        for ℓ in d do
9:            if SQ(d \ {ℓ}) = true then
10:               d ← d \ {ℓ}
11:       φ ← φ ∨ d
12:   return φ
```

The queries performed in Algorithm 3 satisfy the conditions of the transformation: the hypothesis φ is always below the true formula, as required for equivalence queries; the subset queries are always positive adjacent, because if d is a term s.t. $d \implies \psi$, and $d' \not\Rightarrow \psi$ where $d' = d \setminus \{\ell\}$, then taking a state $\sigma^- \models d' \wedge \neg\psi$ and flipping the variable in ℓ results in a state $\sigma^+ \models d$ and hence $\sigma^+ \models \psi$, hence $\sigma^- \models \partial^-(\psi)$ and $\sigma^- \models d'$, as required. As such, the transformation also yields an efficiency result for Algorithm 2 which is carried over from the efficiency of Algorithm 3 for monotone DNF formulas [1,2]:

Theorem 4. *Let $(Init, \delta, Bad)$ be a transition system with $|\Sigma| = n$ and $k \in \mathbb{N}$. If there is an inductive invariant $I \in$ Mon-DNF that is backwards k-fenced, then MB-ITP(Init, [δ], Bad, k) converges to an inductive invariant in $O(m)$ inductiveness queries, $O(mn)$ Hoare queries (with argument k), and $O(mn)$ time.*

Theorm 4 focuses on efficiency of MB-ITP for monotone invariants under the fence condition; in [11] we also show that if *any* k-fenced inductive invariant I exists (not necessarily monotone), then the check in line 5 never fails, hence convergence with reachability bound k is guaranteed. Together with Lemma 1 this provides an alternative proof of termination for MB-ITP.

Efficient Inference of Invariants with a Known Monotone Basis Through Exact Learning. Bshouty [5] investigated exact learning of formulas that are not monotone. To this end, he introduced the *monotone theory*. The idea is that a formula φ can be reconstructed as the conjunction of its *monotoniza-tions*, $\mathcal{M}_b(\varphi)$, w.r.t. elements b in a set B that forms a *monotone basis* for φ; a set of states B is a monotone basis for φ if $\varphi \equiv \bigwedge_{b \in B} \mathcal{M}_b(\varphi)$ (such a set always exists, and is related to CNF representations of φ). Bshouty's Λ-algorithm can efficiently learn φ, while using equivalence and membership queries, provided that the monotone basis is known a-priori, and is amenable to the transformation in Corollary 2, resulting in a Hoare-query algorithm, Λ-Inference, that can *efficiently* learn every formula for which $B = \{b_1, \ldots, b_t\}$ is a basis when the k-fenced condition holds:

Theorem 5. *Let $(Init, \delta, Bad)$ be a transition system with $|\Sigma| = n$, and $k \in \mathbb{N}$. If there exists an inductive invariant I that is backwards k-fenced, $I \in \mathrm{DNF}_m$, and $B = \{b_1, \ldots, b_t\}$ is a monotone basis for I, then Λ-Inference($Init, [\delta], Bad, k$) converges to an inductive invariant in $O(m \cdot t)$ inductiveness checks, $O(m \cdot t \cdot n^2)$ k-BMC checks, and $O(m \cdot t \cdot n^2)$ time.*

Choosing a Monotone Basis. Some important classes of formulas admit a known basis that the algorithm can use. The class of r-*almost-monotone DNF* is the class of DNF formulas with at most r terms which include negative literals. The set of all states with at most r variables assigned *true* is a basis for this class [5]. When $r = O(1)$, the size of this basis is polynomial in $n = |\Sigma|$. Another interesting class with a known base of size polynomial in n is the class of (arbitrary) DNF formulas with $O(\log n)$ terms, although the construction is less elementary [5].

Applying Theorm 5 with the known basis for r-almost-monotone DNF yields:

Corollary 3. *Let $(Init, \delta, Bad)$ be a transition system with $|\Sigma| = n$, $k \in \mathbb{N}$, and $r = O(1)$. If there exists an inductive invariant I that is backwards k-fenced, and I is r-almost-monotone DNF with m terms, then Λ-Inference($Init, [\delta], Bad, k$) with an appropriate basis converges to an inductive invariant in $poly(m \cdot n)$ inductiveness checks, $poly(m \cdot n)$ k-BMC checks, and $poly(m \cdot n)$ time.*

Dual Inference Under the Dual Fence Condition. Safety problems enjoy a duality between the initial states and the bad states: a formula I is an inductive invariant w.r.t. $(Init, \delta, Bad)$ iff the dual formula $\neg I$ is an inductive invariant w.r.t. the dual transition system $(Bad, \delta^{-1}, Init)$. This gives rise to dual algorithms that, given as input $(Init, \delta, Bad)$, infer an invariant for the dual problem and return the dual invariant. Dual algorithms allow us to translate complexity results from the inference of CNF invariants to the inference of DNF invariants and vice versa. Since our results are conditioned upon the backwards-fence condition, we need to dualize it as well:

Definition 9 (Forward k-Fenced). *I is k-forward fenced if every state in $\partial^+(I)$ is reachable from Init in at most k steps, where $\partial^+(I)$ is the inner boundary of I, the set of states $\sigma^+ \models I$ s.t. there is a state $\sigma^- \not\models I$ that differs from σ^+ in exactly one variable.*

Using duality, we derive efficiency results under the k-forward fence condition for antimonotone CNF invariants (from Theorm 4), and for r-almost antimonotone CNF invariants, which are CNF formulas with at most r clauses that include positive literals (from Corollary 3).

5.3   Inference from Two-Sided Fence and Exact Learning

The previous section has provided a translation of exact learning algorithms, but only those that admit certain requirements on their queries. In this section we simulate arbitrary exact learning algorithms (going beyond the requirements in Corollary 2) relying on a *two-sided* fence condition. An important example of such an exact learning algorithm is the CDNF algorithm by Bshouty [5]. The conditions of the transformation in Sect. 5.2 do not hold because this algorithm performs equivalence queries that can return either positive or negative examples. We now show how to implement any membership or equivalence query to the invariant using the two-sided fence condition.

Lemma 5 (Implementing membership queries). *Let* $(Init, \delta, Bad)$ *be a transition system with* $|\Sigma| = n$ *and* I *an (unknown) inductive invariant that is backwards k_1-fenced and forwards k_2-fenced. Then membership queries to I can be implemented in at most n Hoare queries with reachability bound k_1 and n Hoare queries with reachability bound k_2.*[5]

Given a membership query "$\sigma \in I$?", the idea is to choose some known state $\sigma_0 \in Init$, and gradually walk from σ to σ_0, that is, in each step change one variable in σ to match σ_0 and return *true* if $\mathcal{H}^{(k_1)}(\delta, Init, \neg cube(\sigma)) \neq true$, and false if $\mathcal{H}^{(k_1)}(\delta, cube(\sigma), \neg Bad) \neq true$ (otherwise continue the walk). The rational is that if $\sigma \in I$ but it is not reachable from $Init$ in k_1 steps, the walk will eventually hit the inner boundary of I, which is guaranteed to be reachable in k_1 steps—as can be detected using a Hoare query with $k = k_1$—so that the corresponding Hoare query will return such a counterexample trace; similarly, if $\sigma \notin I$ the walk will eventually hit the outer boundary which is guaranteed to reach Bad in k_2 steps and the Hoare query will detect it.

An equivalence query can be implemented by an inductiveness query and a membership query (as was also noted in Sect. 4.2):

Lemma 6 (Implementing equivalence queries). *Let* $(Init, \delta, Bad)$ *be a transition system with* $|\Sigma| = n$, *and* I *an (unknown) inductive invariant that is forwards k_1-fenced and backwards k_2-fenced. Then given θ it is possible to answer whether θ is an inductive invariant, or provide a counterexample σ such that $\sigma \models \theta, \sigma \not\models I$ or $\sigma \not\models \theta, \sigma \models I$, using an inductiveness query, and at most n Hoare queries with bound k_1 and n Hoare queries with bound k_2.*

---

[5] The proof of this also implies that an invariant that is both forwards k_1-fenced and backwards k_2-fenced is unique, seeing that the implementation of the membership query for both is the same.

We can use these procedures to implement every exact learning algorithm from (arbitrary) equivalence and membership queries.

Corollary 4. *Let C be a class of formulas. Let $\mathcal{A}$ be an exact concept learning algorithm that can identify every $\varphi \in C$ in at most s_1 equivalence queries and s_2 membership queries. Then there exists a sound invariant inference algorithm that can find an inductive invariant for every transition system that admits an inductive invariant $I \in C$ that is forwards k_1-fenced and backwards k_2-fenced using at most $s_1 + 1$ inductiveness queries, $n(s_1 + s_2)$ Hoare queries with bound k_1, $n(s_1 + s_2)$ Hoare queries with bound k_2, and time $O(n(s_1 + s_2)t_{\mathcal{A}})$ where $t_{\mathcal{A}}$ is the worst-case time of $\mathcal{A}$ learning I and $n = |\Sigma|$.*

Next, we demonstrate an application of Corollary 4 to the inference of a larger class of invariants.

Inference Beyond Almost-Monotone Invariants. Earlier, we have shown that almost-monotone *DNF* invariants are efficiently inferrable when the *backwards* fence condition holds, and similarly for almost-antimonotone *CNF* when the *forwards* fence condition holds. We now apply Corollary 4 to the CDNF algorithm by Bshouty [5] to show that the class of invariants that can be succinctly expressed *both* in DNF and in CNF (not necessarily in an almost-monotone way) can be efficiently inferred when the fence condition holds in *both* directions:

Theorem 6. *There is an algorithm $\mathcal{A}$ that for every input transition system $(Init, \delta, Bad)$ with $|\Sigma| = n$ and $k \in \mathbb{N}$, if the system admits an inductive invariant I such that $I \in \mathrm{DNF}_{m_1}$, $I \in \mathrm{CNF}_{m_2}$, and I is both backwards- and forwards- k-fenced, then $\mathcal{A}(Init, [\delta], Bad, k)$ converges to an inductive invariant in $O(m_1 \cdot m_2)$ inductiveness queries, $O(m_1 \cdot m_2 \cdot n^3)$ Hoare queries with bound k, and $O(m_1 \cdot m_2 \cdot n^3)$ time.*

Such a complexity guarantee is significant, because, put differently, it shows that an invariant can be learned efficiently in terms of its smallest DNF and CNF representations (provided that the two-sided fence condition holds). Through an observation by Bshouty [5], this implies that it is possible to efficiently infer an invariant that admits a succinct representation as a *decision tree*: a binary tree in which every internal node is labeled by a variable and a leaf by *true/false*, and σ satisfies the formula if the path defined by starting from the root, turning left when the σ assigns *false* to the variable labeling the node and right otherwise, reaches a leaf *true*. The size of a decision tree is the number of leaves in the tree.

Corollary 5. *There is an algorithm $\mathcal{A}$ that for every input transition system $(Init, \delta, Bad)$ with $|\Sigma| = n$ and $k \in \mathbb{N}$, if the system admits an inductive invariant I that can be expressed as a decision tree of size m, and I is both backwards- and forwards- k-fenced, then $\mathcal{A}(Init, [\delta], Bad, k)$ converges to an inductive invariant in $O(m^2)$ inductiveness queries, $O(m^2 \cdot n^3)$ Hoare queries with bound k, and $O(m^2 \cdot n^3)$ time.*

Similarly, when an r-almost-unate invariant with $O(\log n)$ non-unate variables is fenced both backwards and forwards, it can be inferred by an adaptation of an algorithm by Bshouty [6].

Inference Beyond Concept Learning: Efficient Inference of CDNF Invariants from One-Sided Fence Condition. The previous section has arrived at an extremely almost-satisfying result: that any invariant can be inferred in time proportional to the size of its smallest representations in DNF and CNF and the number of variables. The culprit is that the translation from Bshouty's CDNF algorithm is possible only under the *two-sided* fence condition, which is significantly stronger than a one-sided fence condition. In [13] we show that the same result is also attainable under the one-sided fence condition, where the Hoare-query algorithm we use cannot be understood as a direct translation of an exact concept learning algorithm—it builds heavily on Bshouty's CDNF algorithm, but modifies it in important ways, while still accessing the transition relation only through Hoare queries. Specifically, the forwards k-fence condition ensures that the set S of states reachable in at most k steps satisfies $\partial^+(I) \subseteq S \subseteq I$. The algorithm relies on this property to construct a formula H that contains I by sampling and generalizing states from S in a certain way that guarantees that $\partial^+(I) \subseteq H \implies I \subseteq H$. In this way, the algorithm relies on the fact that $\partial^+(I) \subseteq S$ (thanks to the fence condition) to ensure that after sampling enough states from S and using them to increase H, once $\mathcal{H}^{(k)}(\delta, \mathit{Init}, H)=\mathit{true}$ holds (i.e., $S \subseteq H$), then it is also guaranteed that $I \subseteq H$. This process has no analog in exact concept learning, because, there, we are not given any set S that is related to the boundary of the target concept.

6   Conclusion

This paper surveyed results that formally established the relation between SAT-based invariant inference and exact learning with queries, and utilized it to illuminate some of the fundamental questions about invariant inference. There is still much to understand about this topic. In particular, it is interesting to show separation between Hoare queries that use different lengths of executions k, which could indicate that bounded model checking in principle provides additional power. The boundaries of the ability to translate learning algorithms to invariant inference under the fence condition could be clarified by showing that general membership queries are impossible to implement even under the fence condition, justifying the two-sided condition for a general transformation. Finally, other translations that build on reachability conditions other than the fence condition could help explain inference algorithms other than model-based interpolation, and pave the way for new algorithms that are efficient in practice.

Acknowledgement. The research leading to these results has received funding from the European Research Council under the European Union's Horizon 2020 research and innovation programme (grant agreement No. [759102-SVIS]). This research was partially supported by the United States-Israel Binational Science Foundation (BSF) grant No. 2016260, and the Israeli Science Foundation (ISF) grant No. 1810/18.

References

1. Aizenstein, H., Pitt, L.: On the learnability of disjunctive normal form formulas. Mach. Learn. **19**(3), 183–208 (1995). https://doi.org/10.1007/BF00996269
2. Angluin, D.: Queries and concept learning. Mach. Learn. **2**(4), 319–342 (1987)
3. Bjørner, N., Gurfinkel, A., Korovin, K., Lahav, O.: Instantiations, zippers and EPR interpolation. In: LPAR 2013, 19th International Conference on Logic for Programming, Artificial Intelligence and Reasoning, December 12–17, 2013, Stellenbosch, South Africa, Short papers proceedings, pp. 35–41 (2013). https://easychair.org/publications/paper/XtN
4. Bradley, A.R.: Sat-based model checking without unrolling. In: Verification, Model Checking, and Abstract Interpretation - 12th International Conference, VMCAI 2011, Austin, TX, USA, 23–25 January 2011. Proceedings, pp. 70–87 (2011). https://doi.org/10.1007/978-3-642-18275-4_7
5. Bshouty, N.H.: Exact learning Boolean function via the monotone theory. Inf. Comput. **123**(1), 146–153 (1995). https://doi.org/10.1006/inco.1995.1164
6. Bshouty, N.H.: Simple learning algorithms using divide and conquer. Comput. Complex. **6**(2), 174–194 (1997). https://doi.org/10.1007/BF01262930
7. Chockler, H., Ivrii, A., Matsliah, A.: Computing interpolants without proofs. In: Hardware and Software: Verification and Testing - 8th International Haifa Verification Conference, HVC 2012, Haifa, Israel, 6–8 November 2012. Revised Selected Papers, pp. 72–85 (2012). https://doi.org/10.1007/978-3-642-39611-3_12
8. Eén, N., Mishchenko, A., Brayton, R.K.: Efficient implementation of property directed reachability. In: International Conference on Formal Methods in Computer-Aided Design, FMCAD 2011, Austin, TX, USA, October 30–November 02 2011, pp. 125–134 (2011). http://dl.acm.org/citation.cfm?id=2157675
9. Ezudheen, P., Neider, D., D'Souza, D., Garg, P., Madhusudan, P.: Horn-ice learning for synthesizing invariants and contracts. In: PACMPL 2 (OOPSLA), pp. 131:1–131:25 (2018)
10. Feldman, Y.M.Y., Immerman, N., Sagiv, M., Shoham, S.: Complexity and information in invariant inference. In: Proceedings of the ACM Programming Languages, vol. 4 (POPL), pp. 5:1–5:29 (2020). https://doi.org/10.1145/3371073, https://doi.org/10.1145/3371073
11. Feldman, Y.M.Y., Sagiv, M., Shoham, S., Wilcox, J.R.: Learning the boundary of inductive invariants. In: Proceedings of the ACM Programming Languages, vol. 5 (POPL), pp. 1–30 (2021). https://doi.org/10.1145/3434296, https://doi.org/10.1145/3434296
12. Feldman, Y.M.Y., Sagiv, M., Shoham, S., Wilcox, J.R.: Property-directed reachability as abstract interpretation in the monotone theory. In: Proceedings of the ACM Programming Languages, vol. 6 (POPL), pp. 1–31 (2022). https://doi.org/10.1145/3498676, https://doi.org/10.1145/3498676
13. Feldman, Y.M.Y., Shoham, S.: Invariant inference with provable complexity from the monotone theory. In: Static Analysis - 29th International Symposium, SAS 2022, Auckland, New Zealand (To appear in 2022)
14. Flanagan, C., Leino, K.R.M.: Houdini, an annotation assistant for ESC/Java. In: Oliveira, J.N., Zave, P. (eds.) FME 2001. LNCS, vol. 2021, pp. 500–517. Springer, Heidelberg (2001). https://doi.org/10.1007/3-540-45251-6_29

15. Garg, P., Löding, C., Madhusudan, P., Neider, D.: ICE: A robust framework for learning invariants. In: Biere, A., Bloem, R. (eds.) CAV 2014. LNCS, vol. 8559, pp. 69–87. Springer, Cham (2014). https://doi.org/10.1007/978-3-319-08867-9_5

16. Garg, P., Neider, D., Madhusudan, P., Roth, D.: Learning invariants using decision trees and implication counterexamples. In: Proceedings of the 43rd Annual ACM SIGPLAN-SIGACT Symposium on Principles of Programming Languages, POPL 2016, St. Petersburg, FL, USA, 20–22 January 2016, pp. 499–512 (2016). https://doi.org/10.1145/2837614.2837664, https://doi.org/10.1145/2837614.2837664

17. Hellerstein, L., Kletenik, D., Sellie, L., Servedio, R.A.: Tight bounds on proper equivalence query learning of DNF. In: COLT 2012 - The 25th Annual Conference on Learning Theory, 25–27 June 2012, Edinburgh, Scotland, pp. 31.1-31.18 (2012). http://proceedings.mlr.press/v23/hellerstein12/hellerstein12.pdf

18. Henzinger, T.A., Jhala, R., Majumdar, R., McMillan, K.L.: Abstractions from proofs. In: Proceedings of the 31st ACM SIGPLAN-SIGACT Symposium on Principles of Programming Languages, POPL 2004, Venice, Italy, 14–16 January 2004, pp. 232–244 (2004). https://doi.org/10.1145/964001.964021, https://doi.org/10.1145/964001.964021

19. Itzhaky, S., Bjørner, N., Reps, T.W., Sagiv, M., Thakur, A.V.: Property-directed shape analysis. In: Computer Aided Verification - 26th International Conference, CAV 2014, Held as Part of the Vienna Summer of Logic, VSL 2014, Vienna, Austria, 18–22 July 2014. Proceedings, pp. 35–51 (2014). https://doi.org/10.1007/978-3-319-08867-9_3

20. Jha, S., Gulwani, S., Seshia, S.A., Tiwari, A.: Oracle-guided component-based program synthesis. In: Proceedings of the 32nd ACM/IEEE International Conference on Software Engineering - Volume 1, ICSE 2010, Cape Town, South Africa, 1–8 May 2010, pp. 215–224 (2010). https://doi.org/10.1145/1806799.1806833, https://doi.org/10.1145/1806799.1806833

21. Jha, S., Seshia, S.A.: A theory of formal synthesis via inductive learning. Acta Inform. **54**(7), 693–726 (2017). https://doi.org/10.1007/s00236-017-0294-5

22. Jhala, R., McMillan, K.L.: Interpolant-based transition relation approximation. Logical Methods Comput. Sci. **3**(4) (2007). https://doi.org/10.2168/LMCS-3(4:1)2007

23. Koenig, J.R., Padon, O., Immerman, N., Aiken, A.: First-order quantified separators. In: Donaldson, A.F., Torlak, E. (eds.) Proceedings of the 41st ACM SIGPLAN International Conference on Programming Language Design and Implementation, PLDI 2020, London, UK, 15–20 June 2020, pp. 703–717. ACM (2020). https://doi.org/10.1145/3385412.3386018

24. Lahiri, S.K., Qadeer, S.: Complexity and algorithms for monomial and clausal predicate abstraction. In: Schmidt, R.A. (ed.) CADE 2009. LNCS (LNAI), vol. 5663, pp. 214–229. Springer, Heidelberg (2009). https://doi.org/10.1007/978-3-642-02959-2_18

25. McMillan, K.L.: Interpolation and SAT-based model checking. In: Hunt, W.A., Somenzi, F. (eds.) CAV 2003. LNCS, vol. 2725, pp. 1–13. Springer, Heidelberg (2003). https://doi.org/10.1007/978-3-540-45069-6_1

26. McMillan, K.L.: Lazy abstraction with interpolants. In: 18th International Conference on Computer Aided Verification, CAV 2006, Seattle, WA, USA, 17–20 August 2006, Proceedings, pp. 123–136 (2006). https://doi.org/10.1007/11817963_14

27. Neider, D., Madhusudan, P., Saha, S., Garg, P., Park, D.: A learning-based approach to synthesizing invariants for incomplete verification engines. J. Autom. Reason. **64**(7), 1523–1552 (2020). https://doi.org/10.1007/s10817-020-09570-z

28. O'Donnell, R.: Analysis of Boolean Functions. Cambridge University Press, Cambridge (2014). http://www.cambridge.org/de/academic/subjects/computer-science/algorithmics-complexity-computer-algebra-and-computational-g/analysis-boolean-functions
29. Quine, W.: Two theorems about truth-functions. Boletín de la Sociedad Matemática Mexicana **10**(1–2), 64–70 (1954)
30. Reps, T.W., Sagiv, S., Yorsh, G.: Symbolic implementation of the best transformer. In: Verification, Model Checking, and Abstract Interpretation, 5th International Conference, VMCAI 2004, Venice, Italy, 11–13 January 2004, Proceedings, pp. 252–266 (2004). https://doi.org/10.1007/978-3-540-24622-0_21
31. Sharma, R., Aiken, A.: From invariant checking to invariant inference using randomized search. Formal Methods Syst. Des. **48**(3), 235–256 (2016). https://doi.org/10.1007/s10703-016-0248-5
32. Sharma, R., Gupta, S., Hariharan, B., Aiken, A., Liang, P., Nori, A.V.: A data driven approach for algebraic loop invariants. In: Programming Languages and Systems - 22nd European Symposium on Programming, ESOP 2013, Held as Part of the European Joint Conferences on Theory and Practice of Software, ETAPS 2013, Rome, Italy, March 16–24, 2013. Proceedings, pp. 574–592 (2013). https://doi.org/10.1007/978-3-642-37036-6_31
33. Sharma, R., Gupta, S., Hariharan, B., Aiken, A., Nori, A.V.: Verification as learning geometric concepts. In: Logozzo, F., Fähndrich, M. (eds.) SAS 2013. LNCS, vol. 7935, pp. 388–411. Springer, Heidelberg (2013). https://doi.org/10.1007/978-3-642-38856-9_21
34. Sharma, R., Nori, A.V., Aiken, A.: Interpolants as classifiers. In: Computer Aided Verification - 24th International Conference, CAV 2012, Berkeley, CA, USA, 7–13 July 2012 Proceedings, pp. 71–87 (2012). https://doi.org/10.1007/978-3-642-31424-7_11
35. Thakur, A.V., Lal, A., Lim, J., Reps, T.W.: Posthat and all that: automating abstract interpretation. Electr. Notes Theor. Comput. Sci. **311**, 15–32 (2015). https://doi.org/10.1016/j.entcs.2015.02.003
36. Valiant, L.G.: A theory of the learnable. Commun. ACM **27**(11), 1134–1142 (1984). https://doi.org/10.1145/1968.1972, https://doi.org/10.1145/1968.1972
37. Vizel, Y., Grumberg, O.: Interpolation-sequence based model checking. In: Proceedings of 9th International Conference on Formal Methods in Computer-Aided Design, FMCAD 2009, 15–18 November 2009, Austin, Texas, USA, pp. 1–8 (2009). https://doi.org/10.1109/FMCAD.2009.5351148
38. Vizel, Y., Grumberg, O., Shoham, S.: Intertwined forward-backward reachability analysis using interpolants. In: Tools and Algorithms for the Construction and Analysis of Systems - 19th International Conference, TACAS 2013, Held as Part of the European Joint Conferences on Theory and Practice of Software, ETAPS 2013, Rome, Italy, 16–24 March 2013. Proceedings, pp. 308–323 (2013). https://doi.org/10.1007/978-3-642-36742-7_22
39. Vizel, Y., Gurfinkel, A., Shoham, S., Malik, S.: IC3 - flipping the E in ICE. In: 18th International Conference on Verification, Model Checking, and Abstract Interpretation - VMCAI 2017, Paris, France, 15–17 January 2017, Proceedings, pp. 521–538 (2017)

Post's Correspondence Problem: From Computer Science to Algebra

Laura Ciobanu[✉][iD]

Heriot-Watt University and Maxwell Institute, Edinburgh EH14 4AS, Scotland
L.Ciobanu@hw.ac.uk

Abstract. In this short survey we describe recent advances on the Post Correspondence Problem in group theory that were inspired by results in computer science. These algebraic advances can, in return, provide a source of interesting problems in more applied, computational settings.

Post's Correspondence Problem (PCP) is a classical decision problem in theoretical computer science that asks whether for a pair of free monoid morphisms $g, h : \Sigma^* \to \Delta^*$ there is any non-trivial $x \in \Sigma^*$ such that $g(x) = h(x)$. One can similarly phrase a PCP for general groups, rather than free monoids, by asking whether pairs g, h of group homomorphisms agree on any inputs. This leads to interesting and unexpected (un)decidability results for PCP in groups.

Keywords: Post Correspondence Problem · Free and hyperbolic groups · Free monoids · Nilpotent groups · Decidability

1 Introduction

Post's Correspondence Problem (PCP) is a prominent undecidable problem in computer science which owes its popularity both to its particularly simple statement and the fact that it acts as a source of undecidability in a variety of settings. Among the many equivalent ways of formulating PCP we choose the one that lends itself most easily to algebraic manipulation and generalisation: given two finite sets Σ and Δ, PCP takes as input a pair of free monoid morphisms $g, h : \Sigma^* \to \Delta^*$, and asks if there exists any non-trivial $x \in \Sigma^*$ such that $g(x) = h(x)$. Undecidability was proven by Post in 1946 [29], and numerous other problems were proved to be undecidable by reducing them to the PCP: the matrix mortality question, tiling problems, or decidability questions about context-free grammars [17, 27].

One of the successful ways to use the PCP in applications has been to modify it or generalise it. Of the many variations and results on PCP we will survey the following directions:

(1) 'free' PCP: the PCP and its variations for (certain types of) morphisms in free monoids and groups, and
(2) 'beyond free' PCP: the PCP for group homomorphisms in non-free groups.

A. W. Lin et al. (Eds.): RP 2022, LNCS 13608, pp. 28–36, 2022.
https://doi.org/10.1007/978-3-031-19135-0_2

2   'Free' PCP

The PCP in the context of free groups was mentioned for the first time in the 1980's by Stallings [34], to our knowledge, and it remains open for free groups. It is defined analogous to the free monoid case: Let Σ and Δ be two alphabets, let $g, h : F(\Sigma) \to F(\Delta)$ be two homomorphisms from the free group over Σ to the free group over Δ, and store this data in a four-tuple $I = (\Sigma, \Delta, g, h)$, called an *instance* of the PCP. The PCP is the decision problem:

Given $I = (\Sigma, \Delta, g, h)$, is there $x \in F(\Sigma) \setminus \{1\}$ such that $g(x) = h(x)$?

An equivalent way of stating the PCP in free groups is via the *equaliser*

$$Eq(g, h) = \{x \in F(\Sigma) \mid g(x) = h(x)\}$$

of g and h, which is a subgroup of $F(\Sigma)$; the PCP then asks if $Eq(g, h)$ is non-trivial. Equalisers are very natural objects because they are subgroups, and so we can study them with the help of algebraic, combinatorial or geometric tools. Moreover, since $Eq(g, h)$ has an algebraic structure, it makes sense to ask not just about the triviality of $Eq(g, h)$ (i.e. PCP), but for a finite description of it.

Injectivity. A first basic observation about PCP for free groups is that if neither g nor h are injective, then PCP is decidable (by the Lemma 1), which is not the case for free monoids. Recall that if $u, v \in F(\Sigma)$ then we write $[u, v] := u^{-1}v^{-1}uv$ for their *commutator*, and that the *kernel* of a map g (with domain $F(\Sigma)$ as above) is defined as $\ker(g) = \{x \in F(\Sigma) \mid g(x) = 1\}$.

Lemma 1 ([6]). *If $g, h : F(\Sigma) \to F(\Delta)$ are both non-injective homomorphisms then $Eq(g, h)$ is non-trivial.*

Proof. We prove that $\ker(g) \cap \ker(h)$ is non-trivial, which is sufficient. Let $u \in \ker(g)$ and $v \in \ker(h)$ be non-trivial elements. If $\langle u, v \rangle \cong \mathbb{Z} = \langle x \rangle$, there exist integers k, l such that $u = x^k$ and $v = x^l$. Then $g(x^{kl}) = 1 = h(x^{kl})$ so $x^{kl} \in \ker(g) \cap \ker(h)$ with x^{kl} non-trivial, as required. If $\langle u, v \rangle \not\cong \mathbb{Z}$ then $g([u, v]) = 1 = h([u, v])$, so $[u, v] \in \ker(g) \cap \ker(h)$ with $[u, v]$ non-trivial, as required. $\square$

We write $PCP^{(\neg inj, \neg inj)}$ for the PCP where none of the maps is injective, and PCP^{inj} for the PCP with at least one map injective. Lemma 1 settles $PCP^{(\neg inj, \neg inj)}$, leaving PCP^{inj} as the interesting case, when the subgroup $Eq(g, h)$ is finitely generated [15] and a finite description relates to bases (or generating sets): The *Basis Problem* (BP) takes as input an instance $I = (\Sigma, \Delta, g, h)$ of the PCP^{inj} and outputs a basis for $Eq(g, h)$. Recent results settle the BP for certain classes of free group maps [4,8,12,22], as we describe below, but despite this progress its solubility remains open in general. The analogous problem for free monoids aims to describe the equaliser in terms of automata rather than bases, and is insoluble [19, Theorem 5.2].

An interesting subclass of injective maps are the marked morphisms for free monoids, and their analogue, called immersions, for free groups. A set of words $s \subseteq \Delta^*$ is *marked* if any two distinct $u, v \in s$ start with a different letter of Δ, which implies $|s| \leq |\Delta|$. A free monoid morphism $f : \Sigma^* \to \Delta^*$ is *marked* if the set $f(\Sigma)$ is marked. An *immersion of free groups* is a morphism $f : F(\Sigma) \to F(\Delta)$ where the set $f(\Sigma \cup \Sigma^{-1})$ is marked. For these classes of maps not only PCP, but also the stronger Basis Problem, was shown to be decidable.

Theorem 1 ([8,16]). *If g, h are marked morphisms from Σ^* to Δ^*, then one can explicitly compute a finite alphabet $\Sigma_{g,h}$ and a marked morphism $\psi_{g,h} : \Sigma_{g,h}^* \to \Sigma^*$ such that $Image(\psi_{g,h}) = Eq(g, h)$.*

In particular, PCP and the Basis Problem BP are decidable for marked morphisms of free monoids.

Theorem 2 ([8]). *If g, h are immersions from $F(\Sigma)$ to $F(\Delta)$, then one can explicitly compute a finite alphabet $\Sigma_{g,h}$ and an immersion $\psi_{g,h} : \Sigma_{g,h} \to F(\Sigma)$ such that $Image(\psi_{g,h}) = Eq(g, h)$.*

In particular, PCP and BP are decidable for immersions of free groups.

Random Homomorphisms and Generic Behaviour. A different perspective on the PCP and its variations is to consider the behaviour of these problems when the pairs of homomorphisms are picked randomly. Formally: Fix the two alphabets $\Sigma = \{x_1, \ldots, x_m\}$ and $\Delta = \{y_1, \ldots, y_k\}$, $m, k \geq 2$, and ambient free groups $F(\Sigma)$ and $F(\Delta)$, and pick g and h randomly by choosing $(g(x_1), \ldots, g(x_m))$ and $(h(x_1), \ldots, h(x_m))$ independently at random, as tuples of words of length bounded by some positive integer n in $F(\Delta)$. If $\mathcal{P}$ is a property of tuples (or subgroups) of $F(\Delta)$, we say that *generically many* tuples (or finitely generated subgroups) of $F(\Delta)$ satisfy $\mathcal{P}$ if the proportion of m-tuples of words of length $\leq n$ in $F(\Delta)$ which satisfy $\mathcal{P}$ (or generate a subgroup satisfying $\mathcal{P}$), among all possible m-tuples of words of length $\leq n$, tends to 1 when n tends to infinity. There is a vast literature (see for example [18]) on the types of objects and behaviours which appear with probability 1, called *generic*, in infinite groups.

In this spirit, the *generic PCP* refers to the PCP applied to a generic set (of pairs) of maps, that is, a set of measure 1 in the set of all (pairs of) homomorphisms. We say that the generic PCP is decidable if the PCP is decidable for 'almost all' instances, that is, for a set of measure 1 of pairs of homomorphisms.

In [6] we show that among all pairs of homomorphisms g, h, the property of being *conjugacy inequivalent* (that is, for every $u \in F(\Delta)$ there is no $x \neq 1$ in $F(\Sigma)$ such that $g(x) = u^{-1}h(x)u$) occurs with probability 1; that is, conjugacy inequivalent maps are generic. Since for conjugacy inequivalent maps we can choose $u = 1$ we immediately get that if g, h are conjugacy inequivalent then $Eq(g, h) = \{1\}$, which trivially implies the following.

Theorem 3 ([6,7]). *The generic PCP is decidable in free groups.*

The same holds for free monoids, that is, for the classical PCP, by [14].

We collect the results from this section in Table 1.

Table 1. Status of results for free monoids and free groups

Problems	In free monoids	References free monoids	In free groups	References free groups
General PCP	Undecidable	[29]	Unknown	[8]
General BP	Undecidable	[19, Thm 5.2]	Unknown	[8]
$PCP^{(\neg inj, \neg inj)}$	Undecidable	[29]	Decidable	[6]
PCP^{inj}	Undecidable	[23]	Unknown	
PCP marked/immersions	Decidable	[16]	Decidable	[8]
Generic PCP	Decidable	[14, Thm 4.4]	Decidable	[7]

3 'Beyond Free' PCP

We here describe an analogue of PCP for general groups, rather than just for free monoids or free groups, following Levine, Logan and the author [24].

Myasnikov, Nikolaev and Ushakov have previously defined a version of this problem for groups beyond the free ones in [26]. The key difference between their version and ours is how pairs of non-injective maps are dealt with, which we explain below.

Post's Correspondence Problem for Groups. An *instance of the PCP* is a four-tuple $I = (\Sigma, \Gamma, g, h)$ with $g, h : F(\Sigma) \to \Gamma$, where Σ is a finite alphabet and $F(\Sigma)$ is the associated free group, Γ is a group, and $g, h : F(\Sigma) \to \Gamma$ are group homomorphisms. The PCP itself is the decision problem:

Given $I = (\Sigma, \Gamma, g, h)$, is the group $Eq(g, h)/(\ker(g) \cap \ker(h))$ trivial?

Compared to the free group statement in the previous section, where we asked about the triviality of simply $Eq(g, h)$, here we quotient out by $\ker(g) \cap \ker(h)$, as we wish to be able to consider the case when neither map is injective and get substantial information. As we have seen for free groups, Lemma 1 (which easily generalises to non-free groups) immediately deals with the triviality of $Eq(g, h)$ for non-injective maps, and handles PCP, but without getting to the core of the problem. Thus we remove the kernels from the discussion: for non-injective maps $\ker(g) \cap \ker(h)$ is automatically non-trivial as it contains the non-trivial subgroup $[\ker(g), \ker(h)]$, and we end up considering a proper quotient of $Eq(g, h)$.

Note that this definition is completely applicable to free groups as well, and it matches the 'standard' PCP if one of the maps is injective, or if we consider immersions. However, we separated free groups with the standard definition of PCP from more general groups and the 'kernel' definition of PCP because we wanted to couple free groups with free monoids, where kernels do not exist.

By a *solution* to I we mean an element $x \in Eq(g, h) \setminus (\ker(g) \cap \ker(h))$. Solutions are therefore those elements $x \in F(\Sigma)$ that correspond to non-trivial cosets $x(\ker(g) \cap \ker(h)) \in Eq(g, h)/(\ker(g) \cap \ker(h))$.

Myasnikov–Nikolaev–Ushakov considered the triviality of $Eq(g, h)$ and mitigated against $\ker(g) \cap \ker(h)$ automatically being non-trivial by considering varieties of groups. However, this mitigation cannot deal with Γ being hyperbolic, or indeed not being contained in any proper variety of n-generated groups (see Hanna Neumann's book [28] for an exposition on varieties of groups).

Hyperbolic Groups. Hyperbolic groups, and groups of 'negative curvature' more generally, tend to be more manageable when it comes to decision problems and algorithms. The geometry plays a powerful role in limiting the number of computations in the algorithms, and leads to not just decidability, but also easier solutions to algorithmic problems, of lower complexity. For example, in any hyperbolic group the word and conjugacy problem are solvable in linear time, and it is decidable whether a system of equations and inequations has a solution [5,9], and possible to understand the language-theoretic complexity of solution sets of systems of equations and inequations. However, there are exceptions to the decidability results mentioned above, most notably, the subgroup membership problem, which in general undecidable; also, there is no algorithm to compute finite generating sets for intersections of finitely generated subgroups [31]. The subgroup membership problem will be used in the Theorem 4 below.

The first result of [24] regards the PCP for hyperbolic groups; like both the classical PCP in monoids and the subgroup membership problem, it ends up being a negative result. Placing restrictions on the alphabet Σ, group Γ and maps g and h allows one to investigate the boundary between decidability and undecidability. The *binary PCP* is the PCP restricted to those instances $I = (\Sigma, \Gamma, g, h)$ where $|\Sigma| = 2$; the classical (free monoid) binary PCP is decidable [11]. For $\mathfrak{X}$ a class of finitely generated groups, the PCP *for* $\mathfrak{X}$ is the PCP restricted to those instances $I = (\Sigma, \Gamma, g, h)$ where the group Γ is in $\mathfrak{X}$. We can intersect such classes of instances, and so for example can consider the binary PCP for hyperbolic groups.

Theorem 4 ([24]). *The binary PCP for hyperbolic groups is undecidable.*

The proof of Theorem 4 relies on the undecidability of the subgroup membership problem for hyperbolic groups, and on Belegradek and Osin's version of Rips' construction [2]. Belegradek and Osin prove that for every finitely presented group Q and hyperbolic group H there exists a short exact sequence $1 \to N \to \Gamma \to Q \to 1$ such that Γ is hyperbolic and N is a homomorphic image of H. Then H can be chosen carefully so that N has certain desirable properties, and one can pick Q with undecidable word problem, which implies that N has undecidable membership problem. Since the main argument of the proof relies on reducing the triviality of $Eq(g, h)/(\ker(g) \cap \ker(h))$ to the membership in N, this gives the result.

Virtually Nilpotent Groups. On the other side of hyperbolic groups are the so-called 'groups of non-negative curvature', such as nilpotent groups. Finitely generated nilpotent groups have decidable word and conjugacy problems [3]. Much recent work on algorithms in nilpotent groups has been inspired by work in computer science and led to positive results: the compressed word problem [20], the knapsack problem [21] etc. However, unlike in hyperbolic groups, the satisfiability of systems of equations in (free) nilpotent groups is undecidable

[32], and many papers have showed how solving equations in various nilpotent groups [10,13,30,33] reduces to solving Diophantine equations, which fall into the realm of Hilbert's 10th problem and undecidability. In general, arithmetic is much easier to interpret in nilpotent groups than in groups of negative curvature, and it is therefore unsurprising to run into undecidability. On the other hand, and again contrasting with hyperbolic groups, the subgroup membership problem is decidable, and there exists an algorithm to compute generating sets for intersections of finitely generated subgroups [1,25] (many similar positive results extend to polycyclic groups).

From a geometric and algorithmic viewpoint, hyperbolic groups and nilpotent groups are opposites, and so as the PCP is undecidable for hyperbolic groups, we might expect it to be decidable for nilpotent groups. This is indeed the case, with Myasnikov–Nikolaev–Ushakov proving decidability in their above-mentioned paper [26, Theorem 5.8] (their proof actually addresses our definition of the PCP, rather than theirs). Levine, Logan and the author extend this result to virtually nilpotent groups (i.e. groups containing a nilpotent subgroup of finite index).

Theorem 5 ([24]). *The PCP is decidable for finitely generated virtually nilpotent groups.*

In fact, [24] give general sufficient conditions for when PCP is decidable in virtually $\mathfrak{A}$ groups, where $\mathfrak{A}$ a variety of groups (see Hanna Neumann's book [28], for example), results that can extend to more than just virtually nilpotent groups.

4   The Verbal PCP

In this section we consider the version of the PCP defined by Myasnikov–Nikolaev–Ushakov, which is called the "verbal PCP" in [24]. To differentiate, we shall refer to the PCP we defined in Sect. 3 as the *kernel-based* PCP.

The definition we give now is equivalent to Myasnikov–Nikolaev–Ushakov's definition, but has been rephrased to mirror the definition of the kernel-based PCP: An *instance* of the verbal PCP is an instance of the kernel-based PCP, so a four-tuple $I = (\Sigma, \Gamma, g, h)$ with $g, h : F(\Sigma) \rightarrow \Gamma$. Write $V(\Gamma)$ for the maximal verbal subgroup of $F(\Sigma)$ such that any map $F(\Sigma) \rightarrow \Gamma$ factors through $F(\Sigma)/V(\Gamma)$ (i.e. $V(\Gamma)$ is the verbal subgroup generated by the "laws" of Γ, in the sense of Hanna Neumann). The *verbal PCP* itself is the decision problem:

Given $I = (\Sigma, \Gamma, g, h)$, is the group $Eq(g, h)/V(\Gamma)$ trivial?

Note that $V(\Gamma) \leq (\ker(g) \cap \ker(h))$, so for a fixed instance I, the verbal PCP may have solutions when the kernel-based PCP does not.

Hyperbolic Groups. Suppose Γ is non-elementary hyperbolic. Then Γ contains a non-abelian free group, so $V(\Gamma)$ is trivial. Hence, the verbal PCP for non-elementary hyperbolic groups is simply asking if the equaliser $Eq(g, h)$ is trivial. This compares with the kernel-based PCP as follows:

1. If either g or h is injective then $\ker(g) \cap \ker(h) = V(\Gamma)$, as both are trivial. Hence, the verbal PCP and the kernel-based PCP ask the same question and so have identical solution sets. Decidability here is unknown.

2. If both g and h are injective then $\ker(g) \cap \ker(h)$ is non-trivial (as it contains the non-trivial subgroup $[\ker(g), \ker(h)]$). Hence, the verbal PCP necessarily has a solution, and so is trivially decidable. However, by Theorem 4, the kernel-based PCP is undecidable.

Nilpotent Groups. As we mentioned in Sect. 3, Myasnikov–Nikolaev–Ushakov proved the kernel-based PCP for nilpotent groups, rather than the verbal PCP for these groups as their theorem incorrectly states. Levine, Logan and the author, in work in progress, aim to rectify this situation and prove that the verbal PCP is decidable for nilpotent groups.

5 Conclusions

The Post Correspondence Problem is a classical problem in theoretical computer science that crossed over into algebra and generated a number of interesting (un)decidability results in the last few years. There are many ways to phrase the problem in order to obtain a satisfying statement that does not have trivial answers, and we present here two options besides the standard one: the *kernel-based PCP* and the *verbal PCP*.

The crossover from computer science into algebra has begun only recently, and there are undoubtedly many more avenues to explore by translating questions on free monoids, where decades of literature exists, into groups. Once in the world of groups, there are numerous tools to tackle these problems which are not always available for free monoids, tools coming from geometry, topology and algebra. The important question remains whether the progress and ideas from algebra can inform the work in computer science. One of immediate questions is whether PCP in non-free monoids can be studied successfully, especially in monoids with interesting algebraic and geometric properties, such as (certain) hyperbolic or trace monoids.

Finally, the fundamental question asking the decidability of PCP for free groups remains open, and while it is very challenging, progress on the free group on two generators is on the horizon.

References

1. Baumslag, G., Cannonito, F.B., Robinson, D.J., Segal, D.: The algorithmic theory of polycyclic-by-finite groups. J. Algebra **142**(1), 118–149 (1991). https://doi.org/10.1016/0021-8693(91)90221-S, https://doi.org/10.1016/0021-8693(91)90221-S

2. Belegradek, I., Osin, D.: Rips construction and Kazhdan property (T). Groups Geom. Dyn. **2**(1), 1–12 (2008). https://doi.org/10.4171/GGD/29, https://doi.org/10.4171/GGD/29

3. Blackburn, N.: Conjugacy in nilpotent groups. Proc. Amer. Math. Soc. **16**, 143–148 (1965). https://doi.org/10.2307/2034018, https://doi.org/10.2307/2034018

4. Bogopolski, O., Maslakova, O.: An algorithm for finding a basis of the fixed point subgroup of an automorphism of a free group. Internat. J. Algebra Comput. **26**(1), 29–67 (2016). https://doi.org/10.1142/S0218196716500028
5. Ciobanu, L., Elder, M.: The complexity of solution sets to equations in hyperbolic groups. Israel J. Math. **245**(2), 869–920 (2021). https://doi.org/10.1007/s11856-021-2232-z, https://doi-org.ezproxy.st-andrews.ac.uk/10.1007/s11856-021-2232-z
6. Ciobanu, Laura, Logan, Alan D..: Variations on the Post correspondence problem for free groups. In: Moreira, Nelma, Reis, Rogério (eds.) DLT 2021. LNCS, vol. 12811, pp. 90–102. Springer, Cham (2021). https://doi.org/10.1007/978-3-030-81508-0_8
7. Ciobanu, L., Martino, A., Ventura, E.: The generic Hanna Neumann Conjecture and Post Correspondence Problem (2008). http://www-eupm.upc.es/~ventura/ventura/files/31t.pdf
8. Ciobanu, L., Logan, A.D.: The post correspondence problem and equalisers for certain free group and monoid morphisms. In: 47th International Colloquium on Automata, Languages, and Programming (ICALP 2020), pp. 120:1–120:16 (2020)
9. Dahmani, F., Guirardel, V.: Foliations for solving equations in groups: free, virtually free, and hyperbolic groups. J. Topol. **3**(2), 343–404 (2010). https://doi.org/10.1112/jtopol/jtq010, https://doi-org.ezproxy.st-andrews.ac.uk/10.1112/jtopol/jtq010
10. Duchin, M., Liang, H., Shapiro, M.: Equations in nilpotent groups. Proc. Amer. Math. Soc. **143**(11), 4723–4731 (2015). https://doi.org/10.1090/proc/12630, https://doi.org/10.1090/proc/12630
11. Ehrenfeucht, A., Karhumäki, J., Rozenberg, G.: The (generalized) Post correspondence problem with lists consisting of two words is decidable. Theoret. Comput. Sci. **21**(2), 119–144 (1982). https://doi.org/10.1016/0304-3975(89)90080-7
12. Feighn, M., Handel, M.: Algorithmic constructions of relative train track maps and CTs. Groups Geom. Dyn. **12**(3), 1159–1238 (2018). https://doi.org/10.4171/GGD/466
13. Garreta, A., Miasnikov, A., Ovchinnikov, D.: Random nilpotent groups, polycyclic presentations, and Diophantine problems. Groups Complex. Cryptol. **9**(2), 99–115 (2017). https://doi.org/10.1515/gcc-2017-0007, https://doi.org/10.1515/gcc-2017-0007
14. Gilman, R., Miasnikov, A.G., Myasnikov, A.D., Ushakov, A.: Report on generic case complexity (2007). https://arxiv.org/pdf/0707.1364v1.pdf
15. Goldstein, R.Z., Turner, E.C.: Fixed subgroups of homomorphisms of free groups. Bull. London Math. Soc. **18**(5), 468–470 (1986). https://doi.org/10.1112/blms/18.5.468
16. Halava, V., Hirvensalo, M., de Wolf, R.: Marked PCP is decidable. Theoret. Comput. Sci. **255**(1–2), 193–204 (2001). https://doi.org/10.1016/S0304-3975(99)00163-2
17. Harju, T., Karhumäki, J.: Morphisms. In: Handbook of Formal Languages, Vol. 1, pp. 439–510. Springer, Berlin (1997). https://doi.org/10.1007/978-3-642-59136-5
18. Kapovich, I., Myasnikov, A., Schupp, P., Shpilrain, V.: Generic case complexity, decision problems in group theory, and random walks. J. Algebra **264**(2), 665–694 (2003)
19. Karhumäki, J., Saarela, A.: Noneffective regularity of equality languages and bounded delay morphisms. Discrete Math. Theor. Comput. Sci. **12**(4), 9–17 (2010)
20. König, D., Lohrey, M.: Evaluation of circuits over nilpotent and polycyclic groups. Algorithmica **80**(5), 1459–1492 (2018). https://doi.org/10.1007/s00453-017-0343-z, https://doi-org.ezproxy.st-andrews.ac.uk/10.1007/s00453-017-0343-z

21. König, D., Lohrey, M., Zetzsche, G.: Knapsack and subset sum problems in nilpotent, polycyclic, and co-context-free groups. Algebra Comput. Sci. Contemp. Math. **677**, 129–144. American Mathematical Society, Providence, RI (2016). https://doi.org/10.1090/conm/677, https://doi-org.ezproxy.st-andrews.ac.uk/10.1090/conm/677

22. Laura Ciobanu, A.D.L.: Fixed points and stable images of endomorphisms for the free group of rank two. J. Algebra **591**, 538–576 (2022). https://doi.org/10.1007/BF02095993

23. Lecerf, Y.: Récursive insolubilité de l'équation générale de diagonalisation de deux monomorphismes de monoïdes libres $\varphi x = \psi x$. C. R. Acad. Sci. Paris **257**, 2940–2943 (1963)

24. Levine, A., Logan, A.D.: On Post's correspondence problem for hyperbolic and virtually nilpotent groups. **17**, 991–1008, (2022, work in progress)

25. Lo, E.H.: Finding intersections and normalizers in finitely generated nilpotent groups. J. Symbolic Comput. **25**(1), 45–59 (1998). https://doi.org/10.1006/jsco.1997.0166, https://doi.org/10.1006/jsco.1997.0166

26. Myasnikov, A., Nikolaev, A., Ushakov, A.: The Post correspondence problem in groups. J. Group Theory **17**(6), 991–1008 (2014)

27. Neary, T.: Undecidability in binary tag systems and the Post correspondence problem for five pairs of words **30**, 649–661 (2015)

28. Neumann, H.: Varieties of Groups. Springer-Verlag, New York (1967). https://doi.org/10.1007/978-3-642-88599-0

29. Post, E.L.: A variant of a recursively unsolvable problem. Bull. Amer. Math. Soc. **52**, 264–268 (1946). https://doi.org/10.1090/S0002-9904-1946-08555-9

30. Repin, N.N.: Equations with one unknown in nilpotent groups. Mat. Zametki **34**(2), 201–206 (1983)

31. Rips, E.: Subgroups of small cancellation groups. Bull. London Math. Soc. **14**(1), 45–47 (1982). https://doi.org/10.1112/blms/14.1.45, https://doi-org.ezproxy.st-andrews.ac.uk/10.1112/blms/14.1.45

32. Roman'kov, V.A.: Unsolvability of the problem of endomorphic reducibility in free nilpotent groups and in free rings. Algebra i Logika **16**(4), 457–471, 494 (1977)

33. Roman'kov, V.A.: Diophantine questions in the class of finitely generated nilpotent groups. J. Group Theory **19**(3), 497–514 (2016). https://doi.org/10.1515/jgth-2016-0504, https://doi.org/10.1515/jgth-2016-0504

34. Stallings, J.R.: Graphical theory of automorphisms of free groups. In: Combinatorial Group Theory and Topology (Alta, Utah, 1984), Annals of Mathematics Studies, vol. 111, pp. 79–105. Princeton University Press, Princeton (1987). https://doi.org/10.1007/978-1-4612-4372-4

The Past and Future of Embedded Finite Model Theory

Michael Benedikt[✉]

Oxford University, Oxford, UK
michael.benedikt@cs.ox.ac.uk

Abstract. Embedded finite model theory refers to a formalism for describing finite structures over an uninterpreted signature, which sit within an infinite interpreted structures. Some theory was developed in the 1990s and early 2000s, with a focus on the real field. But the theory applies to arbitrary theories, and is relevant to recent development on graph querying and analysis of data-driven programs involving arithmetic.

In this invited paper we review the framework and some of the basic results on it. We also discuss some open questions, along with some work in progress, joint with Ehud Hrushovski.

1 Introduction

A common scenario in symbolic reasoning is one where we have both uninterpreted structure – relation symbols that range over arbitrary interpretations – along with some structure with a fixed interpretation or a heavily-constrained family of interpretations. There are, of course, many formalisms that deal with such situations, depending on what kinds of problems one might look at: data-aware systems [10] and constrained Horn clauses [16], are formalisms of this nature that have been explored in verification.

But there is an older line of work, deriving from database theory and model theory, that takes a different tack on mixing uninterpreted and interpreted structure: *embedded finite model theory*. In this setting we have a target infinite model M in language L, a disjoint finite relational signature S, and we consider formulas over the language $L \cup S$, with the intention that predicates in S range over finite interpretations within M. Formalizing the intended semantics, two formulas in $L \cup S$ are said to be equivalent (modulo M) if they agree in any expansion of M where all finite interpretations of predicates in S have tuples lying in the domain of M. If we fix an L theory T, we can talk about equivalence modulo T, allowing M to range over models of T. Here we will focus on complete theories T. Pairs consisting of a model M for T and a finite interpretation of S within M are referred to as *embedded finite models* (for T) in [5,8,19]. In analogy, we refer to the case where $S = \{P\}$, a single unary predicate, as an *embedded finite subset*. Reasoning about equivalence of $L \cup S$ formulas can be phrased as reasoning about L formulas with free weak second order variables.

A. W. Lin et al. (Eds.): RP 2022, LNCS 13608, pp. 37–46, 2022.
https://doi.org/10.1007/978-3-031-19135-0_3

When we talk about *evaluation* of $L \cup S$ sentences, we will mean that we take an $L \cup S$ formula ϕ along with a finite interpretation S_0 for predicates in S and we want to know if $T \models \phi(S/S_0)$. When T is the complete theory of model M (e.g. the real field), this is the same as asking whether the expansion of M using S_0 for S satisfies ϕ. This evaluation problem can be reduced to reasoning in T, by just substituting S atoms with disjunctions of equalities involving tuples in S_0. However, we will see that there are alternative approaches.

The main goal of this paper and accompanying talk is to re-introduce this topic. We will outline:

- Some highlights of the theory that was developed several decades ago
- Some recent (unpublished) results
- Some open problems in the area

2   Basic Definitions and Some Highlights of Prior Work

Embedded finite model theory developed out of a formalism for spatial data, *constraint databases*, originating in work of Kannelakis, Kuper, and Revesz [18]. They dealt with first order logic over a vocabulary $L \cup S$, where L is an interpreted signature and S is a finite relational signature disjoint from L. The constraint database setting considered interpretations for the relational signature S by definable sets, and focused on the case where T is the complete theory for one of the natural structures over the reals: for example, the real ordered field. We will not talk about constraint databases here, although the more general setting does raise a number of interesting definability questions that go beyond embedded finite model theory: see, for example, [7,15].

Much work in the 1990s dealt with questions of expressiveness of these logics over the reals, and this led to the question of what happens when the interpretation of each relation in S is restricted to be *finite*: that will be our focus here. Of course, we know a lot about the expressiveness of first-order logic over finite structures: for example, we know that there cannot be a first order sentence $\phi(P)$ that holds of a finite set P exactly when the cardinality of P is even. Is the same true for $L \cup \{P\}$ sentences interpreted over embedded finite subsets for a theory T? Or does the ability to use unbounded quantification over the ambient model add to the expressiveness? It obviously depends on T.

Example 1. Consider the case where L is empty and T is the complete theory of an infinite set $M_\emptyset$. Then in $L \cup \{P\}$ we can express statements like:

$$\phi_{1-out} = \text{``there is an } x \text{ that is not in } P\text{''}$$

$$\phi_{2-in,3-out} = \text{``there are two elements in } P \text{ and there are three elements not in } P\text{''}.$$

But note that ϕ_{1-out} is equivalent to True. And $\phi_{2-in,3-out}$ is equivalent to ϕ_{2-in} saying only that P has size at least 2: we can always find the required witnesses outside of P. Here when we say "is equivalent", we mean always equivalent over interpretations of P that are finite.

Hull and Su [17] showed that the phenomenon above is not an accident: for every ϕ as above there is a ϕ' which quantifies only over P, with ϕ' equivalent to ϕ over finite interpretations of P. In this work we call a ϕ' which only quantifies over the "finite part" a *first-order bounded-quantifier formula* or just 1-bounded for short. It easily follows from Hull and Su's result that having quantification over an ambient infinite set does not allow you to express anything new: for example, the parity of a unary relation P cannot be expressed using such quantification. Here we talked about a single unary predicate P, but Hull and Su's result deals with arbitrary finite relational signatures S. The *active domain* of such a structure is the union of the projections of all relations in S. Then [17] shows that, for every ϕ over S, where quantification is over an infinite set, we can find ϕ' equivalent to ϕ that quantifies only over the active domain of the S interpretations: that is, we can pre-process ϕ to quantify only over the finite part. We will talk about bounded-quantifier formulas to mean those that quantify over the active domain of the S structure. Several other terms have been used for these formulas in the past: e.g. "restricted quantifier" "active domain quantifier". The motivation for our terminology 1-bounded should become clearer when we discuss the higher-order case in Sect. 3.

Example 1 deals with a trivial signature and theory. But in the 1990s it was noted that the same phenomenon arises for more involved settings.

Example 2. Consider the case where L consists of a binary relation $<$ and T is the complete theory of a dense linear order. Then for every relational signature S, every $L \cup S$ formula is equivalent to one that quantifies only over the active domain of the S structure.

Intuitively, all elements outside the active domain of S can be grouped into classes, depending on where they sit in the order relative to the active domain of S. We can replace a quantification over an ambient element with a quantification over one of its neighbors in the active domain of S.

We say that a theory T in language L has *restricted quantifier collapse* (RQC) if for every relational signature S, any $L \cup S$ formula is equivalent (over embedded finite models) to a first-order bounded-quantifier formula. RQC implies quantifier elimination, but we will abuse notation by saying that a theory has RQC if some definable extension with quantifier elimination has it. This is equivalent to extending the notion of bounded-quantifier formula to allow arbitrary first-order L formulas in the base case. Then we can restate the early results as saying: the trivial theory and theory of dense linear order have RQC.

The motivating applications we have in mind involve S being a relational database schema [18] or a graph schema [13]. But much of the theoretical phenomena of interest for a given theory is already evident if we restrict to $S = \{P\}$ with P unary. In this case we speak of *embedded finite subsets*.

Interest in RQC and embedded finite model theory in general was spurred by a result due to Van den Bussche, Paredeans, and Van Gucht in [20]: they showed that RQC holds for the real-ordered group. This was later generalized to the real order field [8]. On the other hand, it was noted that theories that are model-theoretically badly-behaved, like full integer arithmetic, do not have RQC: they can express that P has even parity, and indeed can express uncomputable properties of $|P|$.

There are at least two perspectives on RQC. From a practical perspective, evaluation of an $L \cup S$ formula ϕ can be considered as *deciding T-formulas that are parameterized by data*. Any fixed S structure could be inlined into ϕ, giving an L-formula which can be evaluated using whatever decision procedures are available for T (e.g. quantifier elimination). But assuming ϕ is small and the data is huge, such an approach would be problematic. Alternatively, one could apply RQC, if it is applicable. In essence, we are performing the quantifier elimination step prior to knowing the data, resulting in a parameterized quantifier-free formula. With this pre-processing finished, we can inline the S structure and evaluate using any standard technique from databases or finite model theory. This perspective on RQC is emphasized by Basu [3], who referred to collapse results as "uniform quantifier elimination".

Another perspective on these results is model-theoretic. They give us a way of distinguishing theories that are "nice" from those that are not, and we can compare these to other model-theoretic dividing lines. We overview some results in this line below, giving a quick tour of a few model-theoretic properties relevant to RQC. For details, alternative definitions, and further background on these properties see, e.g. [11].

NFCP Theories. A theory T is NFCP (Not the Finite Cover Property) if it satisfies a strong quantitative form of the compactness theorem. For every $\phi(x_1 \ldots x_j, y_1 \ldots y_k)$ there is number n such that for any finite set S of k-tuples in a model of T, if for every subset S_0 of S of size at most n, $M \models \exists \boldsymbol{x} \bigwedge_{s \in S_0} \phi(\boldsymbol{x}, \boldsymbol{s})$, then $M \models \exists \boldsymbol{x} \bigwedge_{s \in S} \phi(\boldsymbol{x}, \boldsymbol{s})$. Examples of NFCP theories include the theory of algebraically closed fields in each characteristic. In particular, the complex field is NFCP.

NFCP theories are inherently *unordered*: they do not admit a linear definable order. In fact, a much stronger statement is true: NFCP theories are *stable*, informally meaning that there is no formula $\phi(\boldsymbol{x}, \boldsymbol{y})$ such that ϕ restricted to arbitrary large sets defines a linear order.

O-Minimal Theories. We now turn to a model theoretic tameness property relevant to *linear ordered* structures. Consider a theory T over L containing a binary relation $< (x, y)$ such that T implies that $<$ is a linear order. Such a T is *o-minimal* if for every $\phi(x, \boldsymbol{y})$ for every model M of T and any $\boldsymbol{c}$ in M, $\{x | M \models \phi(x, \boldsymbol{c})\}$ is a finite union of intervals. The real ordered group, real ordered field, and the real exponential field are all o-minimal [12].

NIP Theories. NIP theories [22] relate to the well-known notion of VC dimension in learning theory. Given $\phi(x_1 \ldots x_j, y_1 \ldots y_k)$, and a finite set of j-tuples S

in a model M we say that S is *shattered* by ϕ if for each subset S_0 of S there is $\boldsymbol{y}_0$ such that $S_0 = \{\boldsymbol{s} \in S | M \models \phi(\boldsymbol{s}, \boldsymbol{y}_0)\}$. A theory T is NIP if for each formula there is a number that bounds the size of a set shattered by ϕ in a model of T. NIP theories include both ordered and highly unordered structures. Specifically, they contain o-minimal structures, Presburger arithmetic, as well as all stable structures, and hence all NFCP structures.

With these definitions in hand, we now present some old results. A natural model-theoretic sufficient condition for RQC over ordered structures involves o-minimal theories:

Theorem 1. *[8] Every o-minimal theory has RQC.*

Belegradek, Stolboushkin and Taitslin [4] showed RQC for an even broader class, what today are known as *distal theories.* We will not give the definition of distal here – see [21]. But it subsumes o-minimal theories and also Presburger arithmetic, while being contained in NIP. In particular, [4] implies that RQC holds for Presburger arithmetic.

On the "unordered side", it is easy to see [14] that NFCP theories are RQC.

Theorem 2. *Every NFCP theory has RQC.*

An easy observation is that NIP is a necessary condition for RQC:

Proposition 1. *[5] If T has RQC, then T is NIP.*

It is easy to identify NIP theories (and even stable theories) that do not have RQC: one example will be given in Example 3 below. Nevertheless it was shown that in NIP theories one cannot express that $|P|$ is even: indeed, one has a variant of RQC for formulas that are *cardinality-invariant*. The rough statement of the result is:

Theorem 3. *[2] In an NIP theory every $L \cup S$ every formula depending only on the isomorphism type of S is equivalent to a bounded-quantifier formula.*

One Paragraph Summary of Past Work. Prior work gave a decent, albeit incomplete, picture of when first-order unbounded quantifiers can be eliminated. If one looks at arbitrary $L \cup S$ formulas, one can make due with either NFCP for unordered theories, or o-minimality/distality for ordered theories. If one focuses on cardinality-invariant formulas, then the broad class of NIP structures is sufficient. Although the topic has attracted sporadic interest within model theory, it seems to us that both the theory and the practice related to it has not been developed as far as one might have hoped. When the motivating application to spatial databases fell out of favor, interest in this topic within the database community faded considerably.

3   Higher Order Boundedness

RQC is restrictive in that it eliminates unbounded first-order quantification, and only allows one to introduce additional first-order bounded quantifiers in return. From the motivation of getting manageable "data complexity" – i.e. as the number of tuples in the embedded finite structures gets large, for fixed ϕ – it might suffice to get an algorithm that scales polynomially in the model; e.g. within fixed point logic. It may even suffice to get an algorithm running in PSPACE within the size of P. This motivates the more general notions of collapse below.

We say that a formula of $L \cup \{P\}$ is k-bounded, for $k > 0$ if it is built up from L-formulas, P stoms and atoms using higher-order variables of order at most k, using the Boolean connectives and quantifications $\exists S$, where S is a higher order variable. The variables in the quantification range over the corresponding set-theoretic hierarchy over P. Thus a 2-bounded formula could use first-order bounded quantification $\exists x \in P \ \phi$ and second-order bounded quantification, such as $\exists S \subseteq P \ \phi$.

A theory is $\forall$ k-bounded for $k > 0$ if every $L \cup \{P\}$ formula is equivalent to a k-bounded one. A theory is $\forall \omega$-bounded if every formula is equivalent to a k-bounded for some k. When $k = 1$ this is just RQC, so we are providing a weaker form of collapse. Above we deal with the case where the relational signature consists of a single unary relation, but there is an obvious extension to arbitrary relational signatures S, where we will quantify over the cumulative hierarchy based on the active domain of the S structure.

The only published result about higher order boundedness is a simple observation that decidable 2-bounded theories exist: (see [19]).

Proposition 2. *The random graph is $\forall$ 2-bounded but not $\forall$ 1-bounded.*

On the negative side, it is quite obvious that really badly-behaved theories do not have this more general form of collapse.

Proposition 3. *The theory of full integer arithmetic $(N, +, \cdot)$ is not ω-bounded.*

As with RQC, these higher-order collapse questions can be seen from an algorithmic perspective, or as a way of classifying theories into tame/wild. From the second perspective, an obvious question is whether the $\forall$ 2-boundedness of the random graph is explained by some model-theoretic property which it satisfies. For example, the theory of the random graph is known to be a *simple theory* [23]. We will not give the definition here, but in addition to the random graph it includes the theory of any pseudo-finite field, where the latter is an ultraproduct of finite fields. Unfortunately, it turns out that standard existing model-theoretic properties do not seem to imply $\forall$ 2-boundedness or even $\forall \omega$-boundedness. For example, "simplicity" does not suffice to imply this more generous notion of boundedness.

Theorem 4. *[6] For each k there are theories that are ω-bounded but not k-bounded. Such theories can be taken to be decidable and simple. There are also decidable simple theories that are not even $\forall$ ω-bounded.*

The construction of examples that are $k+1$ but not k bounded requires the arity of relations in the signature L to grow with k. We do not know of examples in a finite relational signature that are $\forall$ ω-bounded but not $\forall$ k-bounded for some k.

The examples provided by the constructions above are arguably artificial. One natural example of unboundedness comes from pseudo-finite fields mentioned above:

Theorem 5. *[6] For any pseudo finite field in positive characteristic, its theory is not $\forall$ ω-bounded.*

Above we have referred to the complete theory of any pseudo-finite field. The incomplete theory of pseudo-finite fields consists of the sentences true of all finite fields (in a given characteristic). The theory of pseudo-finite fields is known to be decidable [1], but little is known about its complexity. The encoding used in the proof of the theorem above can be used to provide some lower bounds on the theory. And indeed he encoding techniques used, and the connection between non-boundedness results and lower bounds for decidable theories may be the aspect of the theorem above that is most relevant for a theoretical computer science audience.

It is an open question what happens for pseudo-finite fields in characteristic 0. We do not even know if the theory is 2-bounded.

4   Persistent Unboundedness

Like quantifier-elimination, boundedness is sensitive to the signature. It is easy to construct theories that are "badly behaved" – not even ω-bounded – but which become RQC when the theory is expanded to a larger vocabulary.

Example 3. Let $L = \{E(x,y)\}$ and T be the L-theory stating that E is an equivalence relation with classes of each finite size. Consider the $L \cup \{P\}$ formula $\phi_{allclass}$ stating "some equivalence class is contained in P". It is straightforward to show $\phi_{allclass}$ is not equivalent to a k-bounded formula for any k. Informally, no matter how much one knows about the internal structure of P, one cannot tell if it contains an entire equivalence class.

Let T^+ expand T to $L^+ = \{E, < (x,y)\}$, stating that $<$ is a linear order for which each E equivalence class is an interval with endpoints. Then T^+ can be shown to be $\forall$ 1-bounded. This can be shown using distality and the result of [4] mentioned earlier, but it can also be argued directly.

Thus another direction of research is to distinguish theories where we can obtain boundedness by expanding the theory from those where we cannot. Say that a theory is *potentially $\forall$ k-bounded* if it has some expansion that is $\forall$ k-bounded, and similarly for ω-bounded. The above example is thus potentially $\forall$ 1-bounded. If a theory is not potentially $\forall$ ω-bounded we say that it is *persistently unbounded*. Potential unboundedness is thus a further weakening of collapse.

The following easy result, implicit in prior work, provides many examples of persistent unboundedness.

Proposition 4. *If there is an $L \cup P$ ϕ that is cardinality-invariant which is not equivalent to a k-bounded formula for any k over models of T, then T is persistently unbounded.*

That is, if the witness to unboundedness is cardinality-invariant, then expanding the theory will not help. From the proposition one can see that model-theoretically wild theories like $(N, +, \cdot)$ are persistently unbounded.

What is more surprising is that there are decidable theories which do not allow us to express any new cardinality-invariant properties, but which are persistently unbounded. One natural example comes from Boolean algebra:

Theorem 6. *[6] The theory of atomless Boolean algebra is persistently unbounded.*

But for many natural theories that are decidable but not RQC, we do not know if any kind of boundedness can be achieved by expanding the signature. For example, we do not know what happens for Büchi arithmetic, which does not admit RQC [9]. We also do not know the situation when one imposes some of the model-theoretic tameness criteria mentioned previously. For example, can NIP theories be persistently unbounded?

The theorem above also indicates that theories of Boolean algebra are a good place to investigate how to evaluate $L \cup S$ formulas in the absence of any kind of collapse result. We suspect that Büchi arithmetic is also persistently unbounded, which would give another setting in which non-collapse related evaluation techniques should be investigated.

5   Conclusions and Open Issues

Embedded finite model theory focused on the impact of adding unbounded first-order quantification within model checking of first-order formulas on finite structures. Most of the work done in past decades concerned expressiveness of first-order unbounded quantification in different theories. These "collapse results" allow one to conclude a limit on the expressiveness of quantification over the ambient structure. For example, in RQC theories this question reduces to expressibility in classical finite model theory, and so in RQC theories one cannot express queries like parity. This invited paper has also focused on expressiveness, albeit extending the notion of what it means to eliminate unbounded quantification. We have briefly mentioned two such extensions: allowing higher-order quantification and allowing expansion via additional structure.

But we did not mean to imply that the notions of bounded logics defined here and the investigation of collapse questions associated with them are the end of the story.

Computation and Complexity. Even in terms of first-order collapse results, computational aspects – even complexity analysis – have received little attention. Some upper bounds for the real field case can be found in [3]. One can easily see that if one defines RQC in the original way – building up bounded formulas from only quantifier-free formulas in the base case – then the complexity of collapse is lower bounded by the complexity of quantifier elimination for the underlying theory. But as mentioned before, it is natural to build up from arbitrary L-formulas in the base case, which makes RQC specific to the $L \cup S$ setting: it is about pushing unbounded quantifiers past bounded ones. We know of no results on the complexity of this kind of transformation. Neither do we know of serious algorithmic work on evaluation – via RQC or otherwise – even in the context of natural theories such as real linear arithmetic.

Other Source Languages. In defining extended notions of boundedness, we have extended the target language for collapse results, but we have left the source language the same. Recent work [13] has proposed alternative source languages for querying embedded finite models, based on automata. Fragments of the "hybrid second order logics" that combine unbounded first-order quantification with higher-order bounded quantification [8] are another possible source language.

Other Notions of Uniformity. The classification of RQC results as "uniform" or "parameterized" quantifier elimination suggests that other transformations on first-order formulas could be parameterized by data in the same way. For example, model completeness of a theory T can be defined as the ability to transform an arbitrary formula to an existential one. The uniform version would state that an $L \cup S$ formulas can be converted to one with only atomic negation and unbounded existential quantifiers.

We believe that if this topic is re-examined in light of developments in the last two decades, many other new aspects of the theory will emerge.

References

1. Ax, J.: The elemenentary theory of finite field. Ann. Math. **88**(2), 293–271 (1968)
2. Baldwin, J., Benedikt, M.: Stability theory, permutations of indiscernibles, and embedded finite models. Trans. Am. Math. Soc. **352**, 4937–4969 (2000)
3. Basu, S.: New results on quantifier elimination over real closed fields and applications to constraint databases. J. ACM **46**(4), 537–555 (1999)
4. Belegradek, O.V., Stolboushkin, A.P., Taitslin, M.A.: Extended order-generic queries. Ann. Pure Appl. Logic **97**(1–3), 85–125 (1999)
5. Benedikt, M.: Generalizing finite model theory. In: Logic Colloquium 2003, pp. 3–24. Cambridge University Press, London (2006)
6. Benedikt, M., Hrushovski, E.: Embedded finite model theory notes (2022). Manuscript
7. Benedikt, M., Kuijpers, B., Löding, C., Van den Bussche, J., Wilke, T.: A characterization of first-order topological properties of planar spatial data. J. ACM **53**(2), 273–305 (2006)
8. Benedikt, M., Libkin, L.: Relational queries over interpreted structures. J. ACM **47**(4), 644–680 (2000)
9. Benedikt, M., Libkin, L., Schwentick, T., Segoufin, L.: Definable relations and first-order query languages over strings. J. ACM **50**(5), 694–751 (2003)

10. Calvanese, D., De Giacomo, G., Montali, M.: Foundations of data-aware process analysis: a database theory perspective. In: PODS (2013)
11. Chang, C.C., Keisler. H.J.: Model Theory. North-Holland, New York (1990)
12. van den Dries, L.P.D.: Tame Topology and O-minimal Structures. Cambridge University Press, London (1998)
13. Figueira, D., Jez, A., Lin, A.W.: Data path queries over embedded graph databases. In: PODS (2022)
14. Flum, J., Ziegler, M.: Pseudo-finite homogeneïty and saturation. J. Symbol. Logic **64**(4), 1689–1699 (1999)
15. Grohe, M., Segoufin, L.: On first-order topological queries. ACM Trans. Comput. Log. **3**(3), 336–358 (2002)
16. Gurfinkel, A.: Program verification with constrained horn clauses. In: CAV (2022)
17. Hull, R., Su, J.: Domain independence and the relational calculus. Acta Informatica, **31**, 512–524 (1994)
18. Kanellakis, P.C., Kuper, G.M., Revesz, P.Z.: Constraint query languages. J. Comput. Syst. Sci. **51**(1), 26–52 (1995)
19. Libkin, L.: Embedded finite models and constraint databases. In: Finite Model Theory and Its Applications, pp. 257–337. Springer, Heidelberg (2007). https://doi.org/10.1007/3-540-68804-8_5
20. Paredaens, J., Van den Bussche, J., Van Gucht, D.: First-order queries on finite structures over the reals. SIAM J. Comput. **27**(6), 1747–1763 (1998)
21. Simon, P.: Distal and non-distal nip theories. Ann. Pure Appl. Logic **164**(3), 294–318 (2013)
22. Simon, P.: A Guide to NIP Theories. Cambridge University Press, UK (2015)
23. Wagner, P.: Simple theories. Kluwer, Dordrecht (2000)

Regular Papers

Linearization, Model Reduction and Reachability in Nonlinear ODEs

Michele Boreale[✉] and Luisa Collodi

Università di Firenze, Dipartimento di Statistica, Informatica, Applicazioni,
V.le Morgagni 65, 50134 Florence, IT, Italy
michele.boreale@unifi.it

Abstract. In the analysis of nonlinear ordinary differential equations
(ODEs), linear and Taylor approximations are fundamental tools. Such
approximations are generally accurate only in a local sense, that is near a
given expansion point in space or time. We study conditions and methods
to compute linear approximations of nonlinear ODEs that are accurate
also non locally. Relying on Carleman linearization and Krylov projection, our method yields a small, hence tractable linear system that is
shown to produce accurate approximate solutions, under suitable stability conditions. In the general, possibly non stable case, we provide
an algorithm that, given an initial set and a finite time horizon, builds a
tight overapproximation of the reachable states at specified times. Experiments conducted with a proof-of-concept implementation have given
encouraging results.

Keywords: Nonlinear ODEs · Carleman linearization · Krylov spaces ·
Reachability · Stability

1 Introduction

The analysis of systems of nonlinear ordinary differential equations (ODEs) poses
formidable challenges to theoreticians and practitioners. Among the great variety of existing techniques, many concentrate on specific properties, for instance
stability (e.g. via Lyapunov functions [16, Ch.4]) and safety (e.g. via barrier
certificates [14,17,26]). Other techniques focus on computing detailed, effective
descriptions of the set of reachable states over a given time horizon, possibly
taking into account uncertainties on the initial states, see e.g. [2,3,7,8,10,32]
and references therein. These descriptions, variously called reachsets, flowpipes
etc., are typically obtained in a piecewise fashion; that is, by sewing together
local approximations over different regions of the state space and/or time. Here,
we are interested in: (a) conditions and methods by which a single approximation of a system can be computed that can be accurate also non locally; (b)
understanding if such approximations can be leveraged in reachability analysis.

Given a nonlinear system of ODEs in the state variables $x = (x_1, ..., x_n)^T$

$$\dot{x} = f(x_1, ..., x_n) \qquad (1)$$

© The Author(s), under exclusive license to Springer Nature Switzerland AG 2022
A. W. Lin et al. (Eds.): RP 2022, LNCS 13608, pp. 49–66, 2022.
https://doi.org/10.1007/978-3-031-19135-0_4

(see Sect. 2 for precise definitions), approximation can take place either in space, like when linearizing the system's equations around a point $x = x_0$; or in time, like when Taylor expanding the ODE's solution around a time $t = t_0$. With traditional methods, the resulting description will typically exhibit only a limited, *local* accuracy: the quality of the approximation will tend to get very bad as one gets away from $x = x_0$ and/or $t = t_0$.

Our goal is to devise approximations of nonlinear systems that can be accurate also non-locally. In particular, under suitable assumptions, accuracy should remain good over a long, possibly infinite time horizon. In our method, a crucial step in achieving this goal is the computation of a 'small', hence computationally tractable, linear ODE system that approximates (1). In perspective, this linear system might replace the original system not only for the purpose of global reachability analysis, but also for tasks such as runtime verification [23,28].

The proposed technique is related to *Carleman embedding* [5,18], which is used to transform a given nonlinear system like (1) into an *infinite* linear system (Sect. 3). For the purpose of effective computation, this infinite system is truncated at a finite cut off, obtaining a linear system

$$\dot{z} = Az \tag{2}$$

of dimension M, typically with $M \gg n$. The z variables represent the (approximate) evolution of certain functions of the original state variables x, say $z = \alpha(x) = (\alpha_1(x), ..., \alpha_M(x))^T$. The elements of α are chosen in such a way that an observable of interest of the state x, say $g(x)$, can be expressed as a linear combination of them.

In order to achieve dimension reduction, we blend Carleman embedding with Krylov orthogonal projection techniques [27]. Basically, working in the z-coordinate space, we reduce the system (2) via projection of the matrix A onto an appropriate subspace of dimension $m \ll M$, thus obtaining a linear system of dimension m (Sect. 3). Distinctive features of the proposed approximation scheme are the following: (a) the equations of the resulting linear ODE system, while depending on the given observable function g, are independent of the initial state. In fact, the system guarantees an error of $O(t^m)$ near $t = 0$ for *any* given initial condition. Moreover, the reduced linear system can be computed without having to store the whole matrix A, which can be quite large; (b) under suitable stability conditions, error w.r.t. the exact solution is typically small, at least over a finite time horizon (and provably bounded over an infinite time horizon in a special case; Sect. 4).

We leverage these features in an algorithm to compute overapproximations of reachsets of system (1) at specified times over a given finite time horizon (Sect. 5). The algorithm works in the general, not necessarily stable case. The basic idea is to perform advection of the vertices of an initial convex set (polytope), relying on the reduced, linearized system rather than on (1). Similarly to other proposals [10,22,33], compensation of the errors resulting from nonlinearities is reduced to an optimization problem. Experiments conducted with a proof-of-concept implementation of the approximation scheme have given encouraging results

(Sect. 6). Concerning reachsets, we also offer a comparison with state-of-the art reachability tools on a few examples drawn from the literature. A few concluding remarks are reported in the final section (Sect. 7). Due to space limitations, computational details and proofs have been omitted and will be made available in an online full version of the paper.

Related Work. There exists a vast literature on the linearization of nonlinear systems. In particular, techniques based on Carleman embedding [5,18] and the Koopman approach [21] have recently received a renewed attention. Most related to our work and motivations, Jungers and Tabuada [15] have recently proposed a technique for global approximation of nonlinear ODEs by linear ODEs, based on *polyflows*. These are systems that are exactly linearizable via a change of variables. The only systems admitting exact polyflow solutions are those where the set of all Lie derivatives of the state variables w.r.t. the vector field f form a finite-dimensional vector space: hence they can fundamentally be regarded as linear systems in a higher dimensional space. The technique in [15] is based on building polyflows that approximate the original system, using as a basis the Lie derivatives up to some order N; the resulting system plays a role somewhat similar to the truncated Carleman embedding (2). As $N \to +\infty$, the approximation of [15] becomes exact in the interval of convergence of the Taylor expansion of the solution for any given x_0. Note that this is an asymptotic result that does not easily yield concrete bounds for a fixed N. On the contrary, our results provide concrete error bounds for any fixed m and finite time horizon—and also for an *infinite* time horizon under suitable stability assumptions. Systems that are exactly linearizable via polynomial changes of variables are the subject of [6,29,30]. In [6] we have considered Carleman embedding and Krylov-based approximations, essentially from a local point of view. Here, we provide novel analyses of both local and global errors, and exploit them in a reachability algorithm. General error bounds for the truncated Carleman linearization are considered in [4,11]. The time interval of validity of these bounds is quite small, contrary to ours; moreover, in practice they appear to be significantly more conservative than ours. In [12], an efficient reachability analysis algorithm relying on Carleman linearization is presented, limited to the class of weakly nonlinear, dissipative systems. Dimension reduction is not considered in any of [4,11,12].

Our work is also related to the Koopman approach [21], where the system's dynamics are lifted from the state space X to a higher dimensional space of *observables*, smooth functions $X \to \mathbb{R}$. The advantage of doing so is that the dynamics become linear in the space of observables, although the dimension of this space is infinite. In this framework, global analysis techniques have recently emerged, see in particular [20]. Our method too is centered on a basis of observable functions, the aforementioned α. However our goal is quite specific, with an emphasis on finite dimensionality and error bounds that are valid over a prescribed time horizon.

In the field of tools for continuous and hybrid systems, like Flow* [8] and CORA [2], a mix of techniques are adopted including Taylor Models and different forms of linearization [1–3,7,8,19]. Here we are primarily interested in approximations that are accurate for as long as possible, and in connecting them to reachability analysis. As argued empirically in Sect. 6, our approach brings some benefits in terms of accuracy.

2 Preliminaries

For $x = (x_1, ..., x_n)^T$ a vector of state variables, we consider a system of ODEs

$$\dot{x} = f(x) \tag{3}$$

where $f = (f_1, ..., f_n)^T$ is a vector field of locally Lipschitz analytic functions defined on some open subset $\Omega \subseteq \mathbb{R}^n$. For $x_0 \in \Omega$, we let $x(t; x_0)$ be the unique solution of the ODE system with the initial condition $x(0) = x_0$: the unique solution exists and is real analytic in an open interval containing the origin $t = 0$ (Picard-Lindelöf theorem).

For a real analytic function g defined on some open subset of $\mathbb{R}^n$ that includes the trajectories $x(t; x_0)$ for $x_0 \in \Omega$, we will be interested in studying the *observable* of the system (3) via g, that is the function $g \circ x(t; x_0)$.

Recall that $\mathcal{L}_f(g) := \langle \nabla g, f \rangle = \sum_{j=1}^n \frac{\partial g}{\partial x_j} \cdot f_j$ is the Lie derivative of g (w.r.t. f), and $\mathcal{L}_f^{(k)}(g)$ is the k-th Lie derivative, defined inductively by $\mathcal{L}_f^{(k+1)}(g) := \mathcal{L}_f(\mathcal{L}_f^{(k)}(g))$. We shall omit the subscript f whenever it is understood from the context.

3 Linearization and Dimension Reduction

In this section, we first introduce a method of linearization of the system (3) which is strongly related to Carleman embedding [18]. Then discuss how to reduce the dimension of the resulting system with certain, still local, accuracy guarantees. The discussion in this section expands that in [6, Sect. 4].

Carleman Linearization. Generally speaking, one can apply the following method to $g(x(t; x_0))$, for any suitable observable function g, so we will describe the method in terms of such a generic g. As we will see, for the purposes of building approximate solutions, it will be sufficient to apply the results to each of the n identity functions $g = x_i$, for $i = 1, ..., n$ in turn.

Let us fix a set $\mathcal{A} = \{\alpha_1, \alpha_2, ...\}$ of functions $\alpha_i : \mathbb{R}^n \to \mathbb{R}$. For instance $\mathcal{A}$ might be all monomial functions. We assume that, over its domain of definition, the observable function g can be represented in a *unique* way as a linear combination of functions from $\mathcal{A}$ up to a cutoff $M > 0$. In other words, we assume there are unique $v = (\lambda_1, ..., \lambda_M)^T \in \mathbb{R}^M$ and basis vector $\alpha := (\alpha_1, ..., \alpha_M)^T$ such that

$$g = \sum_{i=1}^M \lambda_i \alpha_i = v^T \alpha. \tag{4}$$

Otherwise, all we require from the functions in $\mathcal{A}$ is that they are analytic[1], and that the Lie derivative of each α_i can in turn be expressed as a unique linear combination of elements from $\mathcal{A}$. That is, for each $i \geq 1$, there is a unique sequence of real coefficients a_{ij} $(j \geq 1)$ such that

$$\mathcal{L}(\alpha_i) = \sum_{j \geq 1} a_{ij} \alpha_j. \tag{5}$$

For the sake of simplicity, we shall assume that, for each i, only finitely many coefficients a_{ij} here are nonzero; this assumption is true e.g. for g a polynomial and $\mathcal{A}$ equal to the set of all monomials. We let A denote the $M \times M$ matrix of the coefficients a_{ij} for $1 \leq i, j \leq M$, and B be the $M \times k$ matrix of possibly nonzero elements $b_{i,j} = a_{i,M+j}$; that is, k is chosen large enough to ensure that, for $1 \leq i \leq M$, we have $a_{ij} = 0$ for each $j > M + k$. We let $\psi \triangleq (\alpha_{M+1}, ..., \alpha_{M+k})^T$.

For any fixed initial condition $x_0 \in \Omega$ of the original system (3), we can form the linear system of ODEs and the initial condition described below. Note that for each fixed $x_0 \in \Omega$, $\psi(x(t; x_0)) : I \longrightarrow \mathbb{R}^k$ is a real analytic function of t defined in an interval I containing the origin. This function will in general not be explicitly available, as it depends on the solution $x(t; x_0)$. The *Carleman linearization* (or embedding) of (3) is given by the following linear, non-autonomous system in the variables $z = (z_1, ..., z_M)^T$ and initial condition

$$\dot{z} = Az + B\psi(x(t; x_0)) \tag{6}$$
$$z(0) = \alpha(x_0) =: z_0. \tag{7}$$

The following result is an almost immediate consequence of the existence and uniqueness of the solution of ODEs (Picard-Lindelöf). For a detailed proof, see [6, Th.3].

Theorem 1 (Carleman linearization). *Let $x_0 \in \Omega$. Then $\alpha(x(t; x_0))$ is the unique solution of the system* (6) *with $z(0)$ as in* (7).

Note that we cannot *explicitly* build the system (6), as the function $\psi(x(t; x_0))$ is in general not available – even when ψ and B are available. Moreover, the matrix A itself can in practice be too large to be explicitly generated. Indeed (6) is the starting point to build an approximation, as detailed in the next subsection.

Dimension Reduction via Krylov Projection. Starting from the linearized system (6), we will neglect the "remainder" $\psi(x(t; x_0))$ and then reduce the resulting linear homogeneous system, by projecting A onto an appropriate subspace of $\mathbb{R}^M$ of dimension $m \ll M$. The differential equations of the reduced linear system will depend on g, but not on the initial state x_0. The method is amenable to an "on the fly" implementation, in the following sense: it only requires building the Lie derivatives of g until a prescribed order m. In detail, the order m coincides with

---

[1] This can be weakened to analyticity in some open set containing all the trajectories $x(t; x_0)$ for $x_0 \in \Omega$.

the dimension of the obtained linear system of ODEs and give rise to approximate solutions of (3) that are locally accurate. The behaviour of the global error will be discussed in Sect. 4. Recall that our goal is to approximate a target function $g(x(t; x_0))$. For the sake of notation, we will adopt the following abbreviation for this function:

$$g(t; x_0) := g(x(t; x_0)). \tag{8}$$

Fix $m \geq 1$ and order the elements in $\mathcal{A}$ in such a way the first M functions, $\alpha = (\alpha_1, ..., \alpha_M)^T$ are those appearing in the (unique) decompositions of the Lie derivatives of g from 0 through $m - 1$: that is, for each $j = 0, ..., m - 1$ there is a (unique) vector $u_j \in \mathbb{R}^M$ such that $\mathcal{L}^{(j)}(g) = u_j^T \alpha$; here $u_0 = v$. We assume without loss of generality that $m \leq M$ (typically, $m \ll M$). From (5) and from the definitions of the matrices A, B and of the functions α and ψ, it follows that, componentwise

$$\mathcal{L}(\alpha) = A\alpha + B\psi. \tag{9}$$

From the definition (4) of g and the linearity of $\mathcal{L}(\cdot)$, it follows that $\mathcal{L}(g) = v^T A\alpha + v^T B\psi$. From the assumed uniqueness of the decomposition of Lie derivatives in $\mathcal{A}$, we have that $v^T A = u_1^T$ and $v^T B = 0$ hence

$$\mathcal{L}(g) = v^T A\alpha.$$

Taking the Lie derivative of the above equation, we have $\mathcal{L}^{(2)}(g) = v^T A(A\alpha + B\psi) = v^T A^2\alpha + v^T AB\psi$, where $v^T AB = 0$ again as a consequence of the uniqueness of the decomposition of $\mathcal{L}^{(2)}(g)$ in $\mathcal{A}$. Proceeding similarly for the subsequent derivatives, that is iterating (9) and exploiting the linearity of $\mathcal{L}(\cdot)$ and the uniqueness of the decomposition of $\mathcal{L}^{(j)}(g)$ in $\mathcal{A}$, for $0 \leq j \leq m - 1$, we arrive at the following conclusions.

$$v^T A^j \alpha = \mathcal{L}^{(j)}(g) \quad (0 \leq j \leq m - 1) \tag{10}$$
$$v^T A^{j-1} B = 0 \quad\quad (1 \leq j \leq m - 1). \tag{11}$$

Now, we consider the m-dimensional Krylov space[2] generated by v and A^T, that is the subspace of $\mathbb{R}^M$

$$\mathcal{K}_m := \text{span}\{v, A^T v, (A^T)^2 v, ..., (A^T)^{m-1} v\}. \tag{12}$$

Comparing (12) and (10), we see that $\mathcal{K}_m$ is the subspace of $\mathbb{R}^M$ spanned by the (column) vectors of the coefficients of the Lie derivatives of g from 0 through $m - 1$. Here we assume without loss of generality that $v \neq 0$ and that $\mathcal{K}_m$ has dimension m—that is, m is small enough that the m vectors listed on the right-hand side of (12) are linearly independent. Let $V = [v_1| \cdots |v_m]$ be an orthonormal basis of $\mathcal{K}_m$, represented as a $M \times m$ matrix (see at the end of the section for computational considerations). Now consider the projection of

---

[2] For an introduction to Krylov spaces, see e.g. [27].

A^T onto $\mathcal{K}_m$ and represent it w.r.t. the basis V, in other words we consider the $m \times m$ matrix

$$H_m := V^T A^T V. \tag{13}$$

Given a vector of m distinct state variables $y = (y_1, ..., y_m)^T$, we let the *reduced* linear system derived from (6) and the corresponding initial condition, derived from (7), be defined as:

$$\dot{y} = H_m^T y \tag{14}$$
$$y(0) = V^T z_0 =: y_0.$$

Note that the reduced equations (14) do not depend on $x_0 \in \Omega$. Informally speaking, the solution $y(t; y_0)$ of the reduced system describes the evolution of the vector $\alpha(x(t; x_0))$, projected onto the subspace $\mathcal{K}_m$, in the coordinates of the basis V. Recalling that $v \in \mathbb{R}^M$ is the vector of the coefficients of g with respect to α, as in (4), it is then natural to consider the following approximation of $g(t; x_0)$:

$$\hat{g}(t; x_0) := v^T V y(t; y_0). \tag{15}$$

In fact, we will see that $v_1 = v/\|v\|_2$, while v is orthogonal to v_j for $j > 1$. Hence the above formula can be simplified to

$$\hat{g}(t; x_0) = \|v\|_2\, y_1(t; y_0). \tag{16}$$

In order to study the quality of this approximation, we introduce the error function relative to g

$$\epsilon_g(t; x_0) := g(t; x_0) - \hat{g}(t; x_0). \tag{17}$$

The following result confirms that this error is small near $t = 0$. Indeed, the Taylor expansions of $\hat{g}(t; x_0)$ and $g(t; x_0)$ up to order $m - 1$ coincide: this is a consequence of the fact that the coordinates (in α) of the Lie derivatives of g from 0 to $m - 1$ span $\mathcal{K}_m$.

Theorem 2. *For each $x_0 \in \Omega$, the function $\epsilon_g(t; x_0)$ is $O(t^m)$ around $t = 0$.*

Explicit local bounds of the error function can be obtained from the Taylor theorem with remainder in Lagrange form, assuming we can construct validated enclosures S and E of $x(t; x_0)$ and $y(t; y_0)$, respectively—which for small t is possible by standard techniques, see e.g. [24] and references therein. We state the result in a form suitable for application to reachability analysis, where an initial set X_0 is explicitly considered. Below, we let $\rho_{X_0} :=$ $\inf_{x_0 \in X_0}\{\rho : \rho$ is the radius of convergence of the Taylor series of $\epsilon_g(t; x_0)$ from $t = 0\}$.

Corollary 1. *Consider a set $X_0 \subseteq \Omega$. Fix t s.t. $\rho_{X_0} > t > 0$ and compact sets $S \subseteq \mathbb{R}^n$ and $E \subseteq \mathbb{R}^m$ such that $X_0 \subseteq S$, $V^T X_0 \subseteq E$ and for each $(\tau, x_0) \in$*

$[0, t] \times X_0$ we have $x(\tau; x_0) \in S$ and $y(\tau; y_0) \in E$, where $y_0 = V^T \alpha(x_0)$. Define

$$\gamma^-(t; S, E) := \frac{t^m}{m!} \left(\min_{x \in S} \mathcal{L}^{(m)}(g)(x) - \max_{y \in E} v^T V (H_m^T)^m y \right) \qquad (18)$$

$$\gamma^+(t; S, E) := \frac{t^m}{m!} \left(\max_{x \in S} \mathcal{L}^{(m)}(g)(x) - \min_{y \in E} v^T V (H_m^T)^m y \right). \qquad (19)$$

Then for each $x_0 \in X_0$ and $\tau \in [0, t]$, $\gamma^-(t; S, E) \leq \epsilon_g(\tau; x_0) \leq \gamma^+(t; S, E)$.

There exists a well-known algorithm for the efficient, "on the fly" construction of the matrices V, B, H_m, the Arnoldi iteration [27]. We illustrate our approximation technique with the following example, an instance of the Van der Pol oscillator (VdP, see [31]). This system is used as a benchmark in a number of papers on reachability for nonlinear ODEs.

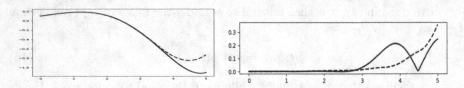

Fig. 1. For Example 1, left: $g(t; x_0)$ and $\hat{g}(t; x_0)$ Example 1; right: $|h(\tau; x_0)|$ and bound (21).

Example 1. Consider the system $\dot{x} = f$ where $f := (x_2, -x_1 + x_2(1 - x_1^2))^T$. We fix as $\mathcal{A}$ the set of all monomials. Let us build the reduced linear system (14) for $m = 2$ and $g = x_1$. This choice of g and m leads to $M = 3$, $\alpha = (x_1, x_2, x_1^2 x_2)^T$, $v = (1, 0, 0)^T$ and $A = \begin{bmatrix} 0 & 1 & 0 \\ -1 & 1 & -1 \\ 0 & 0 & 1 \end{bmatrix}$. Building an orthonormal basis of $\mathcal{K}_m$, we get: $H_m = \begin{bmatrix} 0 & -1 \\ 1 & 1 \end{bmatrix}$ and $V = \begin{bmatrix} 1 & 0 & 0 \\ 0 & 1 & 0 \end{bmatrix}^T$. Writing $x_0 = x = (x_1, x_2)^T$, we have $\hat{g}(t; x) = v^T V y(t; y_0) = y_1(t; y_0)$, where $y(t; y_0)$ is the solution[3] of (14) with initial condition $y_0 := V^T \alpha(x)$. For $x_0 = (0.1, 0.1)^T$, we plot the exact $x_1(t; x_0)$ (dashed) and approximate $\hat{g}(t; x_0)$ (solid) solutions for $t \in [0, 5]$ in Fig. 1, left.

4   Behaviour of the Global Error

We study, mostly from a qualitative point of view, the behaviour of the error function ϵ_g. In what follows, we will assume the orthonormal basis $V = [v_1| \cdots | v_m]$ of the Krylov space $\mathcal{K}_m$ is generated via the Arnoldi Algorithm. This means that the vectors v_j are an orthonormalized version of the vectors $(A^T)^j v$ in (12), inductively built as follows: $v_1 := v / ||v||_2$ and $v_j := w_j / ||w_j||_2$ for $j = 2, ..., m$, where $w_j := A^T v_{j-1} - \sum_{k=1}^{j-1} \mu_k v_k$ with $\mu_k := \langle A^T v_{j-1}, v_k \rangle$. We will let r_m denote the projection of $A^T v_m$ onto $\mathcal{K}_m^\perp$, the orthogonal complement

---

[3] An explicit expression for y_1 is:
  $y_1(t; y_0) = -1/3 \, e^{t/2} (\sqrt{3}(x_1 - 2x_2) \sin(\frac{1}{2} t\sqrt{3}) - 3 x_1 \cos(\frac{1}{2} t\sqrt{3}))$.

of $\mathcal{K}_m$. Explicitly: $r_m := A^T v_m - V V^T A^T v_m$. We define the *remainder function* $h : \mathbb{R}^n \to \mathbb{R}$ as follows for $x \in \mathbb{R}^n$:

$$h(x) := v_m^T B \psi(x) + r_m^T \alpha(x). \tag{20}$$

We have seen that $\hat{g}(t; x)$ represents faithfully $g(t; x)$ up to order $m - 1$ (Theorem 2). Informally speaking, the remainder function h has two error terms, corresponding to whatever of the m-th derivative of $g(t; x)$ cannot be represented: either because it involves elements ψ of $\mathcal{A}$ outside α (term $v_m^T B \psi(x)$), or because it falls outside $\mathcal{K}_m$ (term $r_m^T \alpha(x)$). One's hope here is that $|h(x)|$ is small when computed along the trajectories of $x(t; x_0)$, for x_0 in the initial set. The following theorem provides a general error bound in terms of $h(x)$.

Theorem 3 (global error bound). *For any $t > 0$ such that $x(\tau; x_0)$ is defined for $\tau \in [0, t]$:*

$$|\epsilon_g(t; x_0)| \le ||v||_2 \int_0^t |h(x(\tau; x_0))| \cdot |(e^{(t-\tau) H_m^T})_{1,m}| \, d\tau. \tag{21}$$

If additionally H_m is stable[4] then there is a constant $D > 0$ independent of t such that

$$|\epsilon_g(t; x_0)| \le ||v||_2 D \int_0^t |h(x(\tau; x_0))| \, d\tau. \tag{22}$$

Qualitatively speaking, (22) says that, for a stable H_m, the behaviour of the global error is determined by $|h(x(\tau; x_0))|$: if this function decays fast enough to be integrable over $[0, +\infty)$, then $\epsilon_g(t; x_0)$ will be globally bounded. A special case of this situation (exponential stability) can be easily characterized analytically; details are omitted.

If H_m is not stable, (21) still applies. In this case, the norm of the matrix exponential $e^{(t-\tau) H_m^T}$ will eventually dominate, making the bound useless for large t. Yet, there may be a time horizon within which $|h(x(\tau; x_0))|$ and/or the exponential are small enough to make the bound (21) useful. This will be typically the case if $x(t; x_0)$ hence $|h(x(\tau; x_0))|$ are bounded, for instance in systems that exhibit a limit-cycle behaviour, like VdP. For Example 1, we plot $|h(x(t; x_0))|$ (solid) and the right-hand side of (21) (dashed) for $t \in [0, 5]$ in Fig. 1, right.

These considerations prompt for use of the approximation $\hat{g}(t; x_0)$ inside a scheme for reachability analysis. As for error control, the evaluation of the upper bounds (21) and (22) requires knowledge of the solution $x(t; x_0)$, or at least of a bound on its norm on an interval of interest, which are in general not available. However, useful bounds can be obtained from enclosures of the solution taken at successive, small time intervals. This will be developed in the next section.

---

[4] That is, all eigenvalues of H_m have a nonnegative real part and every imaginary eigenvalue, if any, has geometric multiplicity equal to the algebraic one. See e.g. [16, pp. 135–136].

5   Application to Reachability Analysis

We will apply the linearization scheme outlined in the previous sections to compute an approximation $\hat{x}(t; x_0)$ of the flow $x(t; x_0)$, and then use it to compute an overapproximation of the reachable set of the nonlinear system (1) at fixed times: $t_1, t_2, \ldots$. This goal will be achieved by applying the scheme of Sect. 3 to each of the observable functions $g = x_i$, for $i = 1, \ldots, n$ in turn. Using the notation in that section, for each $i = 1, \ldots, n$, let $v^{(i)}$ the coefficient vector of x_i in the chosen basis α, that is $x_i = v^{(i)T}\alpha$, and $V^{(i)}, H_m^{(i)}$ the corresponding basis and reduced matrix. We define the approximate flow by $\hat{x}(t; x_0) := (\hat{x}_1(t; x_0), \ldots, \hat{x}_n(t; x_0))^T$, where, as an instance of (16), we have

$$\hat{x}_i(t; x_0) := ||v^{(i)}||_2 \, y_1^{(i)}(t; y_0) \quad (i = 1, \ldots, n) \tag{23}$$

with $y^{(i)}(t; y_0)$ the solution of the linear initial value problem (14) for $g = x_i$. Moreover, we will also consider the general case where we are given an initial set X_0 rather than an individual initial state x_0.

The proposed reachability method is inspired by the CHECKMATE algorithm in [10]. For the sake of simplicity, we will represent the initial set X_0 as well as the successive reachsets $R_1, R_2, \ldots$ as convex polytopes[5] (see below). Let $0 = t_0, t_1, \cdots, t_N = t$ be time points, with $\Delta_k := t_k - t_{k-1} > 0$ for $k \geq 1$. The basic idea is to use (approximations of) the *advection maps*

$$x_0 \mapsto x(t_k; x_0) \quad (k = 1, 2, \ldots)$$

to propagate the initial polytope's vertices $\{u_1, \ldots, u_p\}$ to successive time points t_k. At the k-th stage, the polytope resulting from the advected vertices is suitably inflated to compensate for nonlinearities and approximation errors, thus obtaining the actual polytope R_k that over-approximates the reachable set at time t_k; cf. the figure below. The main difference between [10] and us is that, while they approximate the advection maps via numerical integration of the original system (3), we adopt the maps $x_0 \mapsto \hat{x}(t_k; x_0)$. Theorem 3 suggests that $\hat{x}(t_k; x_0)$ is a good approximation of $x(t_k; x_0)$, but does not provide a direct way of bounding the resulting error. Instead, we will keep track of the approximation error via Corollary 1.

Concerning the computation of the approximate advection maps, we recall that, as a solution of the linear system (14), each component in (23) can be written explicitly as:

$$\hat{x}_i(t; x_0) = ||v^{(i)}||_2 \left(e_1^{tH_m^{(i)T}}\right) V^{(i)T}\alpha(x_0) \quad (i = 1, \ldots, n) \tag{24}$$

where $e_1^{(\cdots)}$ denotes the first row of the exponential matrix. As a function of x_0, each $\hat{x}_i(t_k; x_0)$ is a linear combination of the components of the basis $\alpha(x_0)$. For instance, it is a polynomial in x_0 if the elements of the basis are monomials.

---

[5] The method can be extended without much difficulty to more sophisticated and scalable types of sets, like zonotopes.

In more detail, given p vectors (vertices) $u_1, ..., u_p \in \mathbb{R}^n$, we assume that X_0 is the convex hull generated by those vertices, $X_0 = \mathrm{ch}(u_1, ..., u_p) := \{\sum_{i=1}^{p} \lambda_i u_i : \lambda_i \geq 0 \text{ and } \sum_{i=1}^{p} \lambda_i = 1\}$. It is easy to compute the polytope generated by the advected vertices at time t_k, given by $P_k := \mathrm{ch}(\hat{x}(t_k; u_1), ..., \hat{x}(t_k; u_p))$. We let the matrix-vector pair (C_k, b_k) denote a halfspace representation of P_k, that is $P_k = \{x \in \mathbb{R}^n : C_k x \leq b_k\}$. In fact, below we will only make use of the matrix C_k; we assume without loss of generality the rows of this matrix, say $c_1^T, \cdots, c_{\ell_k}^T$, are unitary, $||c_j||_2 = 1$. With the notation used in Corollary 1, we let $\gamma^{-(i)}, \gamma^{+(i)}$ denote the bounds in (18), (19) applied to $g = x_i$, for $i = 1, ..., n$. In Definition 1 below, for $k \geq 1$ we shall adopt the abbreviations

$$\gamma_k^{(i)} := \max\{|\gamma^{-(i)}(\Delta_k; S_k, E_k^{(i)})|, |\gamma^{+(i)}(\Delta_k; S_k, E_k^{(i)})|\} \tag{25}$$

for given compact sets $S_k \supseteq \{x(\tau; \xi) : (\tau, \xi) \in [0, \Delta_k] \times R_{k-1}\}$ and $E_k^{(i)} \supseteq \{y(\tau; \zeta) : (\tau, \zeta) \in [0, \Delta_k] \times V^{(i)T} R_{k-1}\}$. We let $\gamma_k := (\gamma_k^{(1)}, ..., \gamma_k^{(n)})^T$, with $\gamma_0 := (0, ..., 0)^T$. For any nonnegative vector $\zeta \in \mathbb{R}^n$, we will let $[-\zeta, \zeta]$ denote the hyper-rectangle $[-\zeta_1, \zeta_1] \times \cdots \times [-\zeta_n, \zeta_n] \subseteq \mathbb{R}^n$. Below, we assume $\Delta_k < \rho_{R_{k-1}}$ for each $k \geq 1$.

Definition 1 (reachsets R_k). *With the notation introduced above, for $k = 0, 1, 2, ...$ we define the sequence of vectors $\eta_k = (\eta_k^{(1)}, ..., \eta_k^{(\ell_k)})^T \in \mathbb{R}^{\ell_k}$ and of polytopes $R_k \subseteq \mathbb{R}^n$, as follows. $\eta_0 := 0$, $R_0 := X_0$ and, for $k \geq 1$, inductively:*

$$\eta_k^{(j)} := \max_{\substack{\xi \in R_{k-1} \\ \delta \in [-\gamma_k, \gamma_k]}} c_j^T (\hat{x}(\Delta_k; \xi) + \delta) \quad (j = 1, ..., \ell_k) \tag{26}$$

$$R_k := \{x \in \mathbb{R}^n : C_k x \leq \eta_k\}. \tag{27}$$

We note the following important facts about the above definition. (1) Computing η_k requires γ_k, whose computation in turn only requires enclosures S_k and E_k for 'small' flows $x(\tau; \xi)$ and $y(\tau; \zeta)$, for $\tau \in [0, \Delta_k]$ (cf. Corollary 1). (2) In the definition of R_k, one actually modifies a polytope P_k obtained by directly advecting the *initial* X_0 (sort of 'long' advection), not the preceding set R_{k-1}.

The correctness of the method is expressed by the following lemma, which also gives additional guarantees about the enclosure sets[6] S_k.

Lemma 1 (correctness of R_k). *For each $k = 0, 1, ..., N$ and $x_0 \in X_0$, we have $x(t_k; x_0) \in R_k$. Consequently $S_k \supseteq \{x(\tau; x_0) : (\tau, x_0) \in [t_{k-1}, t_k] \times X_0\}$ for each $k \geq 1$.*

The overall workflow of the method is summarized in Algorithm 1, which we christen CKR, for *Carleman-Krylov Reachability*. The timesteps Δ_k, for $k = 1, 2, ..., N$, are such that $\Delta_k = t_k - t_{k-1}$ and $t_N = T$, the time horizon, which is assumed to be in the interval of definition of $x(t; x_0)$ for each $x_0 \in X_0$. In an actual implementation, the timesteps might be chosen adaptively. In the pseudo-code, we use the abbreviation hs(P) to denote a halfspace representation (C, b) of a polytope P.

---

[6] These are useful in case one wants a flowpipe encapsulating the flow $x(t; x_0)$ for all t's in a given interval, not only at specified time points t_k's.

Algorithm 1. CKR

Input: $f = (f_1, ..., f_n)$, vector field in the variables $x = (x_1, ..., x_n)$; $U \subseteq_{\text{fin}} \mathbb{R}^n$
s.t. $X_0 = \text{ch}(U)$; $m \geq 1$, order of approximation; $T \geq 0$, time horizon; $(\Delta_k)_{k=1}^N$,
timesteps.
Output: RL, a list of reachsets.
1: $\alpha :=$ vector of elements of $\mathcal{A}$ to repr. each of $\mathcal{L}_f^{(j)}(x_i)$ $(0 \leq j \leq m - 1, \; 1 \leq i \leq n)$
2: **for** $i = 1, ..., n$ **do**
3:    $v^{(i)} :=$ vector of coefficients of x_i w.r.t. α
4:    $V^{(i)}, H_m^{(i)} := \text{Arnoldi}(f, v^{(i)}, \alpha, m)$ ▷ relies on $u \mapsto A^T u$
5:    $\hat{x}_i := (t, x) \mapsto ||v^{(i)}||_2 \, (e_1^{t H_m^{(i)T}}) V^{(i)T} \alpha(x)$ ▷ Cf. (24). For $t = t_k$ is an adv. map
6: **end for**
7: $R_0 := \text{hs}(\text{ch}(U))$
8: $RL := [R_0]$
9: **for** $k = 1, 2, ..., N$ **do**
10:    $(C_k, b_k) := \text{hs}(\text{ch}(\hat{x}(t_k, U)))$ ▷ k-th advected polytope
11:    $S_k := \text{enclosure}(f, \Delta_k, R_{k-1})$
12:    $E_k^{(i)} := \text{enclosure}((H_m^{(i)})^T, \Delta_k, (V^{(i)})^T R_{k-1})$ $(1 \leq i \leq n)$
13:    $\gamma_k := \text{apply (25) to } \Delta_k, S_k, E_k^{(i)}$ $(1 \leq i \leq n)$
14:    $\eta_k := \text{apply (26) to } R_{k-1}, \gamma_k, A_k$
15:    $R_k := (C_k, \eta_k)$ ▷ k-th reachset
16:    $\text{append}(RL, R_k)$
17: **end for**
18: **return** RL

Example 2. Let us reconsider the VdP system of Example 1. We fix $m = 2$,
$X_0 = [0.1, 0.2] \times [0.1, 0.2]$, $T = \Delta = 0.1$ and a basis α of monomials. The
approximate advection functions $\hat{x}_1$ and $\hat{x}_2$ are computed in the cycle 2–6 of
Algorithm 1: we have already detailed the computation of $\hat{x}_1$ in Example 1;
one proceeds similarly for $\hat{x}_2$. Overall, writing $x_0 = x = (x_1, x_2)^T$, one obtains
$\hat{x}(\Delta; x) := (\hat{x}_1(\Delta; x), \hat{x}_2(\Delta; x))^T$ where for $\Delta = 0.1$:

$$\hat{x}_1(\Delta; x) = 0.99 \cdot x_1 + 0.10 \cdot x_2 \qquad \hat{x}_2(\Delta; x) = -0.10 \cdot x_1^2 \cdot x_2 - 0.10 \cdot x_1 + 1.09 \cdot x_2.$$

Let us see how the first (and only, for this example) reachset R_1 is computed. The
four vertices $U = \{u_1, u_2, u_3, u_4\}$ of X_0 are advected at time T obtaining $U_1 = \{u_1', u_2', u_3', u_4'\} := \{\hat{x}(\Delta; u_1), \hat{x}(\Delta; u_2), \hat{x}(\Delta; u_3), \hat{x}(\Delta; u_4)\}$. For instance, for $u_1 = (0.10, 0.10)^T$, one has $u_1' = \hat{x}(\Delta; u_1) = (0.20, 0.08)^T$. For the convex hull of U_1, a
halfspace representation (C_1, b_1) is computed (step 10); to compensate for errors,
vector b_1 is replaced by a slightly larger[7] (componentwise) η_1, computed via
optimization (step 14), giving rise to the representation $R_1 = (C_1, \eta_1)$ returned
as output.

Remark 1 (computational considerations). Concerning Algorithm 1, a few considerations are in order. (1) The computation of the enclosures $S_k, E_k^{(i)}$ in steps

---

[7] More precisely: $C_1 = \begin{bmatrix} -0.11074 & -0.99546 & 0.99544 & 0.11392 \\ -0.99385 & 0.09513 & -0.09541 & 0.99349 \end{bmatrix}^T$, $b_1 = (-0.11067, -0.10006,$
$0.20011, 0.22134)^T$ and $\eta_1 = (-0.11065, -0.10002, 0.20015, 0.22142)^T$.

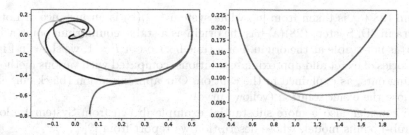

Fig. 2. Individual trajectories starting from $x_0 = (0.485, 0.2)^T$ in the time interval $[0, 1]$ for systems (28)(a) (left) and (28)(b) (right), computed as follows: (i) $x(t; x_0)$, the exact solution computed numerically (yellow); (ii) $\hat{x}(t; x_0)$, our approximate solution (24) (black); (iii) $x_T(t; x_0)$ with $x_{T,i}(t; x_0) := \sum_{j=0}^{m-1} \mathcal{L}^{(j)}(x_i)|_{x=x_0} \frac{t^j}{j!}$, the Taylor expansion of order $m - 1$ of the solution from $t = 0$ (blue), limited to $t = 0.2$ for system (28)(a); (iv) $x_L(t; x_0)$, the solution of the linearized system $\dot{x} = f(x_0) + J_{|x=x_0} \cdot (x - x_0)$, where J is the Jacobian of $f(x)$ in (3) (green). (Color figure online)

11 and 12 can be achieved using any library available to this purpose. We rely on CORA [2] in our implementation. (2) Solving the non-convex optimization problem (26) at step 14 is arguably the most demanding aspect of the algorithm, especially if one is interested in building a certified implementation. In the case of a polynomial vector field and basis, certified upper bounds can be obtained via Sum-Of-Squares (SOS) programming [25], which preserves correctness. In our current implementation, we rely on a general purpose global (non-convex) optimization procedure. We leave for future work the exploration of SOS techniques. (3) Numerical computation of the exponential matrix $e^{(t_k - \tau) H_m^T}$ in step 5 is not problematic, given that m is typically quite small. In a certified implementation, one might compute the exponential via interval arithmetic [13].

6   Experiments

In this section, we present some experimental results obtained by applying the approximation scheme and method in the preceding sections[8]. The section is divided into two parts. First, we compare graphically different approximation methods, including ours, on two examples drawn from the literature, and on a more substantial example drawn from System Biology. Next we illustrate the result of applying a proof-of-concept Python implementation of Algorithm 1, CKR. In particular, we compare CKR with two state-of-the-art tools for reachability analysis, CORA [2] and Flow* [8] on some examples drawn from the literature. In all the examples, for our method we consider a basis of monomial functions.

Graphical Comparisons. We analyze the following two nonlinear systems.

$$(a) \begin{cases} \dot{x}_1 = 4 x_2(x_1 + \sqrt{3}) \\ \dot{x}_2 = -4(x_1 + \sqrt{3})^2 - 4(x_2 + 1)^2 + 16 \end{cases} \quad (b) \begin{cases} \dot{x}_1 = x_1(1.5 - x_2) \\ \dot{x}_2 = -x_2(3 - x_1). \end{cases} \quad (28)$$

---

[8] Code and examples available at https://github.com/Luisa-unifi/CKR.

System (28)(a) is taken from [6], while system (28)(b) is an instance of Lotka-Volterra in 2D. System (28)(a) has the origin as a stable equilibrium point, while (28)(b) is not stable at the origin. For a time horizon of $T = 1$, we show in Fig. 2 trajectories of exact and approximate solutions, computed with various methods, including ours, as explained in the caption. Our approximation (black curve) is very close the exact solution (yellow curve).

We consider also a more substantial example drawn from System Biology, the Laub-Loomis model, whose description we report from [9].

$$\text{(LL)} \begin{cases} \dot{x}_1 = 1.4x_3 - 0.9x_1 & \dot{x}_5 = 0.7x_1 - x_4x_5 \\ \dot{x}_2 = 2.5x_5 - 1.5x_2 & \dot{x}_6 = 0.3x_1 - 3.1x_6 \\ \dot{x}_3 = 0.6x_7 - 0.8x_3x_2 & \dot{x}_7 = 1.8x_6 - 1.5x_7x_2 \\ \dot{x}_4 = 2 - 1.3x_4x_3 \end{cases} \tag{29}$$

Like in the previous example, we make a graphical comparison of the exact solution, computed numerically, with approximate solutions obtained with various methods, including ours. We fix the initial condition to $x_0 = (1.2, 1.05, 1.5, 2.4, 1.0, 0.1, 0.45)^T$ and the time horizon to $T = 1$. The plots in Fig. 3 show that $\hat{x}(t; x_0)$ is quite accurate w.r.t. to the exact solution.

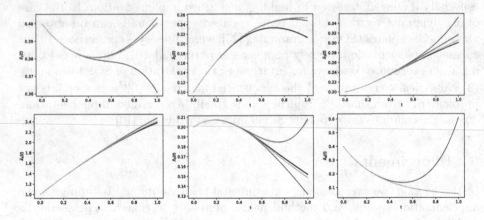

Fig. 3. Exact and approximate solutions for LL over a time horizon of $T = 1$. The color code is the same as in Fig. 2: yellow/exact, black/our approximation, blue/Taylor, green/linearization with jacobian. Different values of m in $\{4, 5, 6\}$ are considered, depending on x_i. For layout reasons, the plot of x_7 is omitted. (Color figure online)

Table 1. Comparison of Flow*, CORA and CKR on: system (28)(b) with $X_0 = [0.40, 0.52] \times [0.18, 0.27]$; system (30) with $X_0 = [-0.5, 0.3] \times [-0.7, 0.8]$; the VdP system of Example 1 with $X_0 = [1.00, 1.50] \times [2.00, 2.45]$. Legenda: **Sys** = system's equation reference, **TH** = time horizon, **Termination** = time at which the algorithm stops, either by natural termination or by breakdown (marked with *), **Accuracy** = average area of reachsets, **m** = approximation order. In each row, the best achieved results are marked in **boldface**.

Sys	TH	Termination Flow* m = 4	m = 8	m = 10	CORA all m	CKR all m	Accuracy (average area) Flow* m = 4	m = 8	m = 10	CORA all m	CKR m = 4/5	Execution time Flow* m = 4	m = 8	m = 10	CORA m = 4	CKR m = 4/5
(28)b	1	1	1	1	1	1	0.02	0.02	0.02	0.01	**0.01**	**0.12**	1.45	3.90	0.17	13.31
	3	3	3	3	2.2*	3	22.75	6.20	6.18	**1.27***	2.82	**0.98**	4.67	14.94	0.47*	50.31
	5	2.7*	5	5	2.2*	5	99.67*	4.37	4.35	1.27*	**1.57**	2.74*	**8.55**	25.24	0.49*	94.79
(30)	1	1	1	1	0.6*	1	5.16	3.34	3.33	5.34*	**0.95**	**0.38**	6.92	23.06	4.28*	14.27
	3	1.3*	1.5*	1.5*	0.6*	3	8.37*	6.81*	6.10*	5.34*	**0.72**	5.04*	21.90*	67.76*	4.18*	**37.96**
	5	1.3*	1.5*	1.5*	0.6*	5	8.37*	6.81*	6.10*	5.34*	**0.62**	4.94*	19.84*	76.48*	5.08*	**64.42**
VdP	1	1	1	1	1	1	0.37	0.37	0.37	0.15	**0.12**	**0.13**	1.71	5.03	2.02	13.72
	3	3	3	3	3	3	0.16	0.15	0.15	0.09	**0.05**	**0.42**	5.05	15.42	5.13	37.05
	5	5	5	5	5	5	0.15	0.13	0.13	0.18	**0.07**	**0.77**	8.54	24.93	10.36	65.66

Reachsets: Comparison with Flow and CORA.* Flow* [8] and CORA [2] are state-of-the-art tools for reachability analysis; they are quite effective at building (over-approximations of) reachsets. The purpose of the following comparison is showing that building reachsets around our the approximate solutions $\hat{x}(t; x_0)$, as we do in CKR, can be beneficial for accuracy. We compare the reachsets R_k produced by CKR with those produced by Flow* and CORA on three examples, one stable and two unstable. Specifically, we consider: (1) the system in (28)(b) (unstable); (2) a new system (stable) defined by:

$$\dot{x}_1 = -x_1^3 + x_2 \qquad \dot{x}_2 = -x_1^3 - x_2^3. \qquad (30)$$

and finally, (3) the VdP system introduced in Example 1 (unstable). VdP also exhibits a limit cycle behaviour. We also stress-test the capabilities of the algorithms in terms of initial sets by considering relatively large X_0's.

We measure the quality of the results as the average area of the reachsets in correspondence of the timesteps returned by each algorithm, until natural or premature termination: $\frac{1}{N} \sum_{k=1}^{N} a(P_k)$, where $a(P_k)$ denotes the area of the polygon P_k corresponding to the reachset at time t_k ($P_k = R_k$ for CKR). We report the obtained results in Table 1, together with the time at which the different algorithms stop, possibly due to an explosion of the overapproximation (breakdown time). For the sake of completeness, we also report a column with execution times[9].

As far as accuracy is concerned, in all cases the sets produced by CKR are tighter than those produced by the other two, often significantly so. Concerning termination, CKR is the only algorithm to complete its execution over the whole time horizon in all the considered cases.

---

[9] It should be noted, though, that it makes little sense to compare a proof-of-concept implementation with highly optimized tools in this respect. At any rate, all execution times are below 100 seconds.

7 Conclusion

We have presented an approach to effectively compute, given a nonlinear ODE system, a linear system which is at the same time small and useful to produce globally accurate approximate solutions, under suitable conditions. We have argued that the method can also bring some benefit to classical reachability analysis in terms of accuracy.

As for future work, we would like to investigate the relation of our method with other well-known linearization schemes, in particular the Koopman approach. Indeed, a trait d'union between Koopman's and our approach is the central role played in both by the observable functions g. In our case, the decomposition of g in the given basis α is the starting point to build a Krylov space, which seems to capture a lot of relevant information about $g(x(t; x_0))$. This is reminiscent of finite-dimensional approximations in Koopman's approach [21, Ch.1.4], but establishing a precise relation is nontrivial.

References

1. Althoff, M.: Reachability analysis of nonlinear systems using conservative polynomialization and non-convex sets. In: Proceedings of the 16th International Conference on Hybrid Systems: Computation and Control (HSCC 2013), pp. 173–182. ACM (2013). https://doi.org/10.1145/2461328.2461358
2. Althoff, M.: An introduction to CORA 2015. In: Proceedings of the Workshop on Applied Verification for Continuous and Hybrid Systems, pp. 120–151 (2015)
3. Althoff, M., Frehse, G., Girard, A.: Set propagation techniques for reachability analysis. Ann. Rev. Control Robot. Auton. Syst. 4(1) (2021). https://doi.org/10.1146/annurev-control-071420-081941. hal-03048155
4. Amini, A., Sun, Q., Motee, N.: Error bounds for Carleman linearization of general nonlinear systems. In: 2021 Proceedings of the Conference on Control and Its Applications. SIAM (2021)
5. Bellman, R., Richardson, J.M.: On some questions arising in the approximate solution of nonlinear differential equations. Q. Appl. Math. 20(4), 333–339 (1963)
6. Boreale, M.: Algorithms for exact and approximate linear abstractions of polynomial continuous systems. In: Proceedings of the 21st International Conference on Hybrid Systems: Computation and Control (Part of CPS Week), HSCC 2018. ACM (2018)
7. Chen, X., Abraham, E., Sankaranarayanan, S.: Taylor model flowpipe construction for non-linear hybrid systems. In: Real-Time Systems Symposium (RTSS), pp. 183–192. IEEE (2012)
8. Chen, X., Ábrahám, E., Sankaranarayanan, S.: Flow*: an analyzer for non-linear hybrid systems. In: Sharygina, N., Veith, H. (eds.) CAV 2013. LNCS, vol. 8044, pp. 258–263. Springer, Heidelberg (2013). https://doi.org/10.1007/978-3-642-39799-8_18
9. Chen, X., Abraham, E., Sankaranarayanan, S.: FLOW* benchmarks. https://flowstar.org/benchmarks/
10. Chutinan, A., Krogh, B.H.: Computational techniques for hybrid system verification. IEEE Trans. Autom. Control 48(1), 64–75 (2003)

11. Forets, M., Pouly, A.: Explicit error bounds for Carleman linearization. arXiv:1711.02552 (2017)
12. Forets, M., Schilling, C.: Reachability of weakly nonlinear systems using Carleman linearization. In: Bell, P.C., Totzke, P., Potapov, I. (eds.) RP 2021. LNCS, vol. 13035, pp. 85–99. Springer, Cham (2021). https://doi.org/10.1007/978-3-030-89716-1_6
13. Goldsztejn, A., Neumaier, A.: On the exponentiation of interval matrices (2009). hal-00411330v1
14. Goubault, E., Jourdan, J.-H., Putot, S., Sankaranarayanan, S.: Finding non-polynomial positive invariants and Lyapunov functions for polynomial systems through Darboux polynomials. In: ACC 2014, pp. 3571–3578 (2014)
15. Jungers, R.M., Tabuada, P.: Non-local linearization of nonlinear differential equations via polyflows. In: 2019 American Control Conference (ACC), pp. 1–6 (2019). https://doi.org/10.23919/ACC.2019.8814337
16. Khalil, H.: Nonlinear Systems, vol. 3/e. Prentice-Hall (2002)
17. Kong, H., He, F., Song, X., Hung, W.N.N., Gu, M.: Exponential-condition-based barrier certificate generation for safety verification of hybrid systems. In: Sharygina, N., Veith, H. (eds.) CAV 2013. LNCS, vol. 8044, pp. 242–257. Springer, Heidelberg (2013). https://doi.org/10.1007/978-3-642-39799-8_17
18. Kowalski, K., Steeb, W.-H.: Nonlinear Dynamical Systems and Carleman Linearization. World Scientific (1991)
19. Makino, K., Berz, M.: Rigorous integration of flows and ODEs using Taylor models. In: Proceedings of Symbolic-Numeric Computation, pp. 79–84. ACM (2009). https://doi.org/10.1145/1577190.1577206
20. Mauroy, A., Mezic, I.: Global stability analysis using the eigenfunctions of the Koopman operator. IEEE Trans. Autom. Control 61(11), 3356–3369 (2016). https://doi.org/10.1109/TAC.2016.2518918
21. Mauroy, A., Mezic, I., Susuki, Y. (eds.): The Koopman Operator in Systems and Control: Concepts, Methodologies, and Applications. Springer, Cham (2020). https://doi.org/10.1007/978-3-030-35713-9
22. Mitchell, I.M., Bayen, A.M., Tomlin, C.J.: A time-dependent Hamilton-Jacobi formulation of reachable sets for continuous dynamic games. IEEE Trans. Autom. Control 50, 947–957 (2005)
23. Mitsch, S., Platzer, A.: ModelPlex: verified runtime validation of verified cyber-physical system models. Formal Methods Syst. Des. 49, 33–74 (2016). https://doi.org/10.1007/s10703-016-0241-z
24. Nedialkov, N.S., Jackson, K.R., Corliss, G.F.: Validated solutions of initial value problems for ordinary differential equations. Appl. Math. Comput. 105(1), 21–68 (1999)
25. Prajna, S., Papachristodoulou, A., Seiler, P., Parrilo, P.: SOSTOOLS and its control applications. In: Henrion, D., Garulli, A. (eds.) Positive Polynomials in Control. LNCIS, vol. 312, pp. 273–292. Springer, Heidelberg (2005). https://doi.org/10.1007/10997703_14
26. Prajna, S.: Barrier certificates for nonlinear model validation. In: 42nd IEEE International Conference on Decision and Control (IEEE Cat. No. 03CH37475), vol. 3, pp. 2884–2889 (2003). https://doi.org/10.1109/CDC.2003.1273063
27. Saad, Y.: Iterative Methods for Sparse Linear Systems. SIAM (2003)
28. Sánchez, C., et al.: A survey of challenges for runtime verification from advanced application domains (beyond software). Formal Methods Syst. Des. 54, 275–335 (2019). https://doi.org/10.1007/s10703-019-00337-w

29. Sankaranarayanan, S.: Automatic abstraction of non-linear systems using change of bases transformations. In: HSCC 2011, pp. 143–152 (2011)

30. Sankaranarayanan, S.: Change-of-bases abstractions for non-linear systems. CoRR abs/1204.4347 (2012)

31. van der Pol, B.: The nonlinear theory of electric oscillations. Proc. Inst. Radio Eng. **22**, 1051–1086 (1934)

32. Tiwari, A., Khanna, G.: Nonlinear systems: approximating reach sets. In: Alur, R., Pappas, G.J. (eds.) HSCC 2004. LNCS, vol. 2993, pp. 600–614. Springer, Heidelberg (2004). https://doi.org/10.1007/978-3-540-24743-2_40

33. Xue, B., Fränzle, M., Zhan, N.: Under-approximating reach sets for polynomial continuous systems. In: Proceedings of the 21st International Conference on Hybrid Systems: Computation and Control (Part of CPS Week), HSCC 2018, pp. 51–60. ACM (2018). https://doi.org/10.1145/3178126.3178133

History-Deterministic Timed Automata
Are Not Determinizable

Sougata Bose[1]([✉]), Thomas A. Henzinger[2], Karoliina Lehtinen[3,4],
Sven Schewe[1], and Patrick Totzke[1]

[1] University of Liverpool, Liverpool, UK
{sougata,sven.schewe,totzke}@liverpool.ac.uk
[2] IST Austria, Klosterneuburg, Austria
tah@ist.ac.at
[3] CNRS, Aix-Marseille University, Marseille, France
[4] LIS, University of Toulon, Toulon, France
lehtinen@lis-lab.fr

Abstract. An automaton is history-deterministic (HD) if one can safely
resolve its non-deterministic choices on the fly. In a recent paper, Henzinger, Lehtinen and Totzke studied this in the context of Timed Automata [9], where it was conjectured that the class of timed ω-languages
recognised by HD-timed automata strictly extends that of deterministic
ones. We provide a proof for this fact.

Keywords: Timed automata · History-determinism

1 Introduction

History-determinism asks for the existence of a strategy to produce a run on
an input word that is given one letter at a time, so that the resulting run is
accepting whenever the given word is in the language.

Similar to automata with bounded ambiguity, history-deterministic ones provide a middle ground between determinism and non-determinism. They are typically more succinct, or even more expressive, than their deterministic counter-
parts while preserving some of their good algorithmic properties. For example,
when verifying finite or ω-automata against history-deterministic specifications
(i.e. checking inclusion with languages given by a HD automaton), the costly step
of complementing the specification automaton can be avoided, as checking language inclusion can be replaced by checking fair simulation [9], which is polynomial for finite, Büchi and co-Büchi automata [8]. For some co-Büchi-recognisable
languages, history-deterministic automata can be exponentially more succinct
than any equivalent deterministic automaton [12], and checking if a Büchi or co-
Büchi automaton is history-deterministic is decidable in polynomial time [2,12].

History-determinism has mostly been studied in the ω-regular setting, i.e., for
finite-state automata recognising languages of infinite words or trees, where the
concept of history-determinism has also been called "good-for-games" [3,6,10,

A. W. Lin et al. (Eds.): RP 2022, LNCS 13608, pp. 67–76, 2022.
https://doi.org/10.1007/978-3-031-19135-0_5

13]. Recently, the notion has been extended to richer automata models, such as pushdown automata [7,14] and quantitative automata [4,5], where deterministic and nondeterministic models have different expressivity.

In [9], history-determinism was first studied in the context of *timed* automata(TA) with ω-regular acceptance conditions [1]. It is shown that for history-deterministic TA with arbitrary parity acceptance, timed universality, inclusion, and synthesis are EXPTIME-complete, contrary to their undecidability for non-deterministic Büchi TA [1]. History-deterministic TA with safety acceptance are effectively determinisable; checking if a given timed automaton is history-deterministic is decidable in EXPTIME for safety or reachability acceptance, and open for more general acceptance conditions such as Büchi, coBüchi and Parity.

In terms of expressivity, it was conjectured that history-deterministic timed automata recognise a class of ω-languages strictly between those defined by deterministic and non-deterministic TAs. The following language is proposed as a candidate to separate deterministic and HD timed languages.

Let L be the language of all timed words along which *eventually* events appear at unit distances: from some time t onwards, for every nonnegative integer i, there is an event at time $t + i$.

It is not difficult to see that this language is recognised by a HD coBüchi automaton. One can commit to the fractional time at which the longest chain of events has been observed so far, and can afford to be wrong a finite number of times. It is intuitively clear that L is not deterministic, considering that any DTA has only finitely many clocks and thus "cannot remember unboundedly many past timestamps" for comparisons. It is however notoriously technical to provide formal arguments for showing that timed languages are not deterministic. Herrmann [11] suggests some high-level lemmas based on reset points, but these only apply to the Büchi setting.

The main contribution of this paper is a formal argument that the language L is indeed not recognised by any deterministic timed automaton, even with general Parity acceptance. We present a scheme to recursively produce, for a given DTA, a suitable pair of words so that their runs are region-equivalent (and so either both are accepting or both rejecting) but where only one of them is in L. The main idea is to produce events and observe the resulting run until it closes a loop in the region graph, then force that same loop again twice more. Any resets that occurred in the intermediate loop are lost and overwritten in the final iteration, which allows to move the timing of the intermediate loop arbitrarily.

We also provide an example that separates history-deterministic from non-deterministic timed automata, concluding that indeed, this class of timed languages sits strictly in between deterministic and non-deterministic ones.

2   Notations

Let $\mathbb{N}$ and $\mathbb{R}_{\geq 0}$ denote the nonnegative integers and reals, respectively. For $c \in \mathbb{R}_{\geq 0}$ we write $\lfloor c \rfloor$ for its integer and $fract(c) \stackrel{\text{def}}{=} c - \lfloor c \rfloor$ for its fractional part.

Timed Alphabets and Words. A timed word is a finite or infinite word $w = (a_0, t_0)(a_1, t_1), \ldots$ over the alphabet $(\Sigma \times \mathbb{R}_{\geq 0})$ where the first components are letters from some finite alphabet Σ and the second components are non-decreasing and progressing, that is, for all $n \in \mathbb{N}$, there is some i and a such that $w[i] = (a, t)$ and $t > n$. We sometimes call the (a_i, t_i) an a_i-*event* with timestamp t_i. For convenience, we will confuse timed words as above with words over $(\Sigma \cup \mathbb{R}_{\geq 0})$, interpreting each letter either as discrete event or a delay. The *duration* of a (finite or infinite) timed word is the combined sum of all its delays. More precisely, a timed word w as above gives rise to the word $d_0 a_0 d_1 a_1 \ldots$ over $(\Sigma \cup \mathbb{R}_{\geq 0})$, where $t_0 = d_0$ and $t_{i+1} = t_i + d_{i+1}$. Conversely, the duration and the timed word of a word over $(\Sigma \cup \mathbb{R}_{\geq 0})$ is given inductively as follows. For any $d \in \mathbb{R}_{\geq 0}$, $a \in \Sigma$, $\alpha \in (\Sigma \cup \mathbb{R})^*$, and $\beta \in (\Sigma \cup \mathbb{R})^\infty$ let $duration(\tau) \stackrel{\text{def}}{=} 0$; $duration(d) \stackrel{\text{def}}{=} d$; $duration(\alpha\beta) = duration(\alpha) + duration(\beta)$; $tword(\varepsilon) = tword(d) \stackrel{\text{def}}{=} \varepsilon$; $tword(\alpha d) \stackrel{\text{def}}{=} tword(\alpha)$; and $tword(\alpha\tau) \stackrel{\text{def}}{=} tword(\alpha)(\tau, duration(\alpha))$.

Timed Automata are finite-state automata equipped with finitely many real-valued variables called *clocks*, whose transitions are guarded by constraints on clocks. Constraints on clocks $C = \{x, y, \ldots\}$ are (in)equalities $x \triangleleft n$ where $x \in C$, $n \in \mathbb{N}$ and $\triangleleft \in \{\leq, <\}$. Let $\mathcal{B}(C)$ denote the set of Boolean combinations of clock constraints, called *guards*. A clock *valuation* $\nu \in \mathbb{R}^C_{\geq 0}$ assigns a value $\nu(x)$ to each clock $x \in C$. We write $\nu \models g$ if ν satisfies the guard g. A timed automaton (TA) $\mathcal{T} = (Q, \iota, C, \Delta, \Sigma, Acc)$ is given by

- Q a finite set of states including an initial state ι;
- Σ an input alphabet;
- C a finite set of clocks;
- $\Delta \subseteq Q \times \mathcal{B}(C) \times \Sigma \times 2^C \times Q$ a finite set of transitions; each transition is associated with a guard, a letter, and a set of clocks to reset.
- $Acc \subseteq \Delta^\omega$ an acceptance condition.

We assume that for all $(s, \nu, a) \in Q \times \mathbb{R}^C_{\geq 0} \times \Sigma$ there is at least one transition $(s, g, a, R, s') \in \Delta$ so that ν satisfies g. $\mathcal{T}$ is *deterministic* if there is at most one such enabled transition. I.e., for every state s and for every letter $a \in \Sigma$, all transitions from s on a have mutually exclusive guards.

A *configuration* is a pair consisting of a control state and a clock valuation. For every configuration $(s, \nu) \in Q \times \mathbb{R}^C_{\geq 0}$,

1. there is a *delay* step $(s, \nu) \stackrel{d}{\longrightarrow} (s, \nu + d)$ for every $d \geq 0$, which increments all clocks by d.
2. there is a *discrete* step $(s, \nu) \stackrel{\tau}{\longrightarrow} (s', \nu')$ if $\tau = (s, g, a, R, s') \in \Delta$ is a transition so that ν satisfies g and $\nu' = \nu[R \to 0]$, that is, it maps all clocks in R to 0 and agrees with ν on all other values.

A path $\rho = (s_0, \nu_0) \stackrel{l_1}{\longrightarrow} (s_1, \nu_1) \stackrel{l_2}{\longrightarrow} (s_2, \nu_2) \ldots$ is called a *run on* timed word $w \in (\Sigma \times \mathbb{R}_{\geq 0})^\infty$ if $tword(l_1 l_2 \ldots) = w$, where $tword(\tau) = a$, for $\tau = (s, g, a, R, s') \in \Delta$. It is *accepting* if $\rho \in Acc$. The language $L(s, \nu) \subseteq (\Sigma \times \mathbb{R}_{\geq 0})^\omega$ of a configuration (s, ν) consists of all timed words for which there exists an accepting run from (s, ν). The language of $\mathcal{T}$ is $L(\mathcal{T}) \overset{\text{def}}{=} L((\iota, 0))$, the language of the initial configuration with state ι and the valuation 0 where all clocks set to zero. We assume that Acc is determined by a parity condition: $Q \to D$ maps states to a integer priority domain $D = [i..p]$ with minimal priority $i \in \{0, 1\}$ and maximal priority $p \in \mathbb{N}$. A run is accepting if the highest priority seen infinitely often along the run is even. Büchi acceptance corresponds to $D = \{1, 2\}$ and co-Büchi acceptance corresponds to $D = \{0, 1\}$.

Regions. The following is the standard definition of regions (cf. [1], def. 4.3). Let $\mathcal{T} = (Q, \iota, C, \Delta, \Sigma, Acc)$ be a timed automaton and for any clock $x \in C$ let c_x denote the largest constant in any clock constraint involving x. Two valuations $\nu, \nu' \in \mathbb{R}_{\geq 0}^C$ are *(region) equivalent* (write $\nu \sim \nu'$) if all of the following hold.

1. For all $x \in C$ either $\lfloor \nu(x) \rfloor = \lfloor \nu'(x) \rfloor$ or both $\nu(x)$ and $\nu'(x)$ are greater than c_x.
2. For all $x, y \in C$ with $\nu(x) \leq c_x$ and $\nu(y) \leq c_y$, $fract(\nu(x)) \leq fract(\nu(y))$ iff $fract(\nu'(x)) \leq fract(\nu'(y))$.
3. For all $x \in C$ with $\nu(x) \leq c_x$, $fract(\nu(x)) = 0$ iff $fract(\nu'(x)) = 0$.

It follows that there are only finitely many equivalence classes w.r.t. $\sim$, called regions, for any given TA. Two configurations (s, ν) and (s', ν') are *(region) equivalent*, write $(s, \nu) \sim (s', \nu')$, if $s = s'$ and $\nu \sim \nu'$. Two runs are *(region) equivalent* if they have the same length and stepwise visit region equivalent configurations. Let $maxfrac(\nu) = \max\{fract(\nu(x)) \mid x \in C\}$ denote the maximal fractional value of any clock in configuration ν. We will make use of the following two properties.

Proposition 1. *1. For any valuation ν and $d \leq 1 - maxfrac(\nu)$ we have $\nu \sim \nu + d$.*

2. Suppose that $(p, \nu) \sim (p', \nu')$ and let $\rho \in (\Delta \cup \mathbb{R}_{\geq 0})^*$ satisfy $duration(\rho) < 1 - maxfrac(\nu)$, $duration(\rho) < 1 - maxfrac(\nu')$ and $(p, \nu) \overset{\rho}{\longrightarrow} (q, \mu)$.
Then $(p', \nu') \overset{\rho}{\longrightarrow} (q', \mu')$ for some $(q', \mu') \sim (q, \mu)$.

Proof (sketch). Part 1 is immediate from the definition of regions.

Part 2 can be shown by induction on the length of ρ using the facts that region-equivalent configurations enable the same discrete transitions and that any delay decreases the duration of the remaining path by the same amount it increases clocks. □

History-Deterministic TA. A timed automaton is history-deterministic if one can resolve non-deterministic choices on-the-fly.

More formally, consider a function $r : (\Delta \cup \mathbb{R}_{\geq 0})^* \times \Sigma \to \Delta$ that, given a finite run $\rho_i = (s_0, \nu_0) \overset{a_0}{\longrightarrow} (s_1, \nu_1) \overset{a_1}{\longrightarrow} (s_2, \nu_2) \ldots (s_i, \nu_i)$ and a next letter $a_i \in \Sigma$, returns a transition $r(\rho_i, a_i) = (s_i, g_i, a_i, s_{i+1}) \in \Delta$ such that $\nu_i \models g$. This yields, for every word $w = a_0 a_1 \ldots \in (\Sigma \cup \mathbb{R}_{\geq 0})^\omega$ and initial configuration (s_0, ν_0), a unique run in which the ith step $(s_i, \nu_i) \overset{a_i}{\longrightarrow} (s_{i+1}, \nu_{i+1})$ either results from a

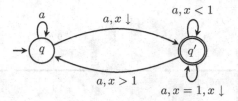

Fig. 1. A history-deterministic timed co-Büchi automaton for L. The state q' has priority 0, i.e. is accepting, while the state q has priority 1.

delay (if $a_i \in \mathbb{R}_{\geq 0}$ and $(s_{i+1}, \nu_{i+1}) = (s_i, \nu_i + a_i)$) or a discrete step chosen by r (if $a_i \in \Sigma$ and $r(\rho_i, a_i) = (s_i, g_i, a_i, s_{i+1})$).

Such a function is called *resolver* if for any input word $w \in L(T)$ the constructed run ρ from initial configuration $(s_0, \nu_0) = (\iota, 0)$ is accepting. A timed automaton is *history-deterministic* if such a resolver exists.

3   D < HD

The interesting aspect of our claim is to show that HD timed automata are strictly more expressive than deterministic ones. We show that the following language L over the singleton alphabet $\Sigma = \{a\}$ is recognised by a one-clock history-deterministic co-Büchi automaton yet not by any deterministic Parity timed automaton. In words, L asks to eventually see events a at unit distance. Formally,

$$L \stackrel{\text{def}}{=} \{(a, t_0)(a, t_1)... \mid \exists i \in \mathbb{N}. \quad \forall n \in \mathbb{N}. \quad \exists j > i. \quad t_j - t_i = n\}.$$

L is HD Recognisable

We show that L is recognised by the history-deterministic timed ω-automaton, in Fig. 1. This automaton has an initial rejecting state q, from where there is a nondeterministic choice to either remain in this state or transition to an accepting state q', which resets the unique clock. There are two transitions to stay in the accepting state: one enabled when the clock value is smaller than 1, and one enabled at clock value 1, which also resets the clock. If the clock value grows larger than 1, the only enabled transition goes back to the initial state. Since this is a co-Büchi automaton, an accepting run must eventually remain in the accepting state.

First, this automaton recognises L: if $w \in L$ then there is an accepting run that moves to state q' at time t, where it then remains since the clock x is reset at the occurrence of each event $(a, t + n)$ for $n \in \mathbb{N}$, so the clock value never grows larger than 1. Conversely, a word accepted by this automaton has a run that eventually moves to q' at a time t, and then remains in q'. For the run to

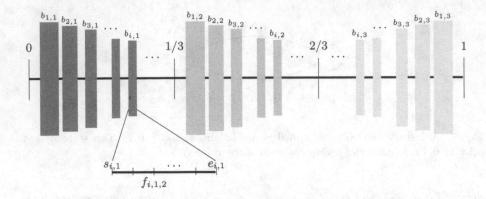

Fig. 2. Blocks within an interval and ticks within a block

stay in q', it must reset x at every time-unit after t, so $(a, t + n)$ must occur in the word for all $n \in \mathbb{N}$, that is, the word is in L.

We now argue that this automaton is also history-deterministic. Given a finite word read so far and a new letter a at time t_{new}, the resolver identifies the earliest time t_{early} such that a has so far occurred at time $t_{early} + n$ for all integers n such that $t_{early} + n \leq t_{new}$. Let r be the function that maps a run ρ ending in q to q' if $t_{new} = t_{early} + m$ for some integer m, and otherwise to the only other available transition.

We claim that this is indeed a resolver. If $w \in L$ then there is an earliest time t such that $(a, t + n)$ occurs in w for all integers n. Since t is minimal, eventually the resolver r will make its choice whether to move to q' over a letter (a, t_{new}) based on whether $t_{new} = t + m$ for some integer m. Since time progresses and $(a, t + n)$ occurs in w for all integers n, the run will eventually transition to q' at a time $t + m$ for some m. From there, since $(a, t + n)$ occurs in w for all integers n, the run over w remains in q' and is therefore accepting.

It remains to be shown that L is not recognised by a deterministic timed automaton.

L is not Deterministic Parity Recognisable

Suppose towards a contradiction that L is recognisable by some deterministic Timed Automaton D with Parity acceptance. Let r be the number of its regions.

We will construct two words, one belonging to L and one that does not, so that the run of D on w is region equivalent to the one on w'. The two words can only differ in the timing of events since there is only one letter in the alphabet.

Both words will be constructed on the fly, according to the following schema.

Consider the intervals and fractional values in Fig. 2; There are infinitely many disjoint intervals, $b_{i,j} = [s_{i,j}, e_{i,j}]$ so that all $b_{i,1}$ have start and endpoint strictly between 0 and $\frac{1}{3}$ and are increasing, i.e., $s_{i+1,1} < e_{i,1}$ for all i. Similarly, $b_{i,2} \subseteq [\frac{1}{3}, \frac{2}{3}]$, and $s_{i+1,2} < e_{i,2}$ for all i. The third sequence of intervals $b_{i,3} \subseteq [\frac{2}{3}, 1]$ have start and endpoint strictly between $\frac{1}{3}$ and 1 and are *decreasing*: $e_{i+1,3} <$

$s_{i,3}$ for all i. Each interval $b_{i,j}$ contains equi-distant values $f_{i,j,0}, f_{i,j,1}, \ldots, f_{i,j,r}$ starting at $f_{i,j,0} = s_{i,0}$.

We step-wise construct w (and w') together with the run of D on it. In every integral interval from $i - 1$ to i we place events as follows.

- start with a delay of $f_{i,1,1}$, followed by a discrete event a, then delay of $f_{i,1,2} - f_{i,1,1}$ followed by a, and so on. This induces a run of D on the pre-fix constructed and we continue constructing the prefix until the induced run closes a cycle in the region graph. This implies existence of times $f_{i,1,k}$ and $f_{i,1,k+\ell}$ such that the automaton is in configurations $(s_{i,1,k}, \nu_{i,1,k})$ and $(s_{i,1,k+\ell}, \nu_{i,1,k+\ell})$ and $(s_{i,1,k+\ell}, \nu_{i,1,k+\ell}) \sim (s_{i,1,k}, \nu_{i,1,k})$. We denote by L_i the run between $f_{i,1,k}$ and $f_{i,1,k+\ell}$.
- Now we force the automaton to close the same cycle, but with all events occurring at times in the interval $b_{i,2}$ (respectively $b_{1,2}$) in w (respectively w'). This can be done by adding a time delay by $s_{i,2} - f_{i,1,k+\ell}$ in w followed by an event a at times $f_{i,2,\ell'}$ for all $\ell' \leq \ell$. We prove this formally in Lemma 1.
- Finally we force the automaton to close the same cycle once more, with all times in interval $b_{i,3}$. This can be done by adding a time delay $s_{i,3} - f_{i,2,\ell}$ followed by events at times $f_{i,3,1}, f_{i,3,2}, \ldots f_{i,3,\ell}$. We prove the correctness of the construction in Lemma 1.

Consider the cycle L_i in the region graph obtained in step 1 above in the interval $[i - 1, i]$, between $f_{i,1,k}$ and $f_{i,1,k+l}$. Note that the k and ℓ depends on i. However, we write k and ℓ without as we only reason about loops within an integral interval. The duration of the loop, denoted by $duration(L_i)$ is $f_{i,1,k+\ell} - f_{i,1,k}$. An important observation is that $duration(L_i) \leq e_{i,j} - s_{i,j}$ as the loop occurs within the interval between $s_{i,1}$ and $e_{i,1}$.

Lemma 1. *Let ν_i and ν'_i be the configurations reached by the run of D at times $i - 1 + f_{i,1,k}$ and $i - 1 + f_{i,1,k+\ell}$. Then $1 - maxfrac(\nu_i + d_{ij}) \geq duration(L_i)$, where $d_{ij} = s_{i,j} - f_{i,1,k}$ for $j \in \{2, 3\}$.*

Furthermore, let ν_{ij} be the configuration reached by the run of D at time $i - 1 + f_{i,j,1}$, where $j = \{2, 3\}$. The cycle L_i is executable from ν_{ij}.

Proof. We prove this lemma by induction on i. The case $i = 1$ is easy to see since $maxfrac(\nu_1 + d_{1j}) \leq s_{1,j}$ and therefore $1 - maxfrac(\nu_1 + d_{1j}) \geq 1 - s_{1,j} \geq e_{1,j} - s_{1,j} \geq duration(L_1)$.

Furthermore, $\nu_{12} = \nu'_1 + d$, where $d = s_{1,2} - f_{1,1,k+\ell} \leq 1 - f_{1,1,k+\ell} \leq 1 - maxfrac(\nu'_1)$. Therefore, by Proposition 1.1., $\nu_{12} \sim \nu'_1 \sim \nu_1$. For $\nu = \nu_1$ and $\nu = \nu_{12}$, $1 - maxfrac(\nu) > e_{1,3}$, $s_{1,3} > duration(L_1)$ as $1 > e_{1,3}$ and $maxfrac(\nu) < s_{1,3}$. By applying Proposition 1.2., L_1 is executable from ν_{12} and ends in a configuration $\nu'_{12} \sim \nu_{12}$.

The configuration ν_{13} equals $\nu'_{12} + d'$, where $d' = s_{1,3} - f_{1,2,\ell} < 1 - maxfrac(\nu'_{12})$ as $maxfrac(\nu'_{12}) \leq f_{1,2,\ell}$. Proposition 1.1. gives $\nu_{13} \sim \nu_{12}$, and $1 - maxfrac(\nu_{13}) \geq e_{1,3} - s_{1,3} \geq duration(L_1)$. By Proposition 1.2., we can conclude that L_1 is executable from ν_{13}.

To prove the inductive case, we bound the value of $maxfrac(\nu_i + d_{ij})$ for $j \in \{2, 3\}$. Consider a clock $x \in C$ and the last time when it was reset. Either it was never reset or the reset occurred at time $f_{i',j',k'}$. For a clock that is never reset, the fractional part of its value at ν_i will be $f_{i,1,k}$. If the clock was last reset within some blue block, i.e., at time $i' - 1 + f_{i',1,k'}$, then either $i' < i$ (corresponds to previous blue blocks), or $k' < k$ (corresponds to previous ticks within the current blue block). In both cases, the $fract(x) = fract(f_{i,1,k} - (i' - 1 + f_{i',1,k'})) \leq f_{i,1,k}$.

Note that any reset to clock x in a previous red block must also be reset again in the corresponding green block as the runs in the red and green block are the same by construction. For a clock x last reset in some previous green block, i.e., at time $i' - 1 + f_{i',3,k'}$, $fract(x) = fract((i - 1 + f_{i,1,k}) - (i' - 1 + f_{i',3,k'})) = f_{i,1,k} + (1 - f_{i',3,k'})$. Furthermore, $f_{i',3,k'} > s_{i-1,3}$ as $i' \leq i$. Therefore, $fract(x) \leq 1 + f_{i,1,k} - s_{i-1,3} + 1$ which bounds $1 - fract(x) \geq s_{i-1,3} - f_{i,1,k}$. Combining all the possibilities for clock resets, we obtain $1 - maxfrac(\nu_i) \geq s_{i-1,3} - f_{i,1,k}$.

It is easy to see that for $j \in \{2, 3\}$, $maxfrac(\nu_i + d_{ij}) \leq maxfrac(\nu_i) + d_{ij}$. Therefore, $1 - maxfrac(\nu_i + d_{ij}) \geq 1 - maxfrac(\nu_i) - d_{ij} \geq 1 - (f_{i,1,k} - s_{i-1,3} + 1) - (s_{i,j} - f_{i,1,k}) \geq s_{i-1,3} - s_{i,j} \geq e_{i,j} - s_{i,2}$. The last step follows from the fact that $e_{i,j} \leq s_{i-1,3}$ for $j \in \{2, 3\}$. Note that the duration of the loop L_i is less than $e_{i,j} - s_{i,j}$ and thus completes the proof for fist part of the lemma.

We now show that L_i is executable from ν_{i2} and ν_{i3}. First, $\nu_{i2} \sim \nu_i$ and $\nu_{i3} \sim \nu_{i2}$ by repeated application of Proposition 1.1.. This is similar to the argument in the base case. We just showed that $1 - maxfrac(\nu_{i2}) > s_{i-1,3} - s_{i,2} > e_{i,2} - s_{i,2} = duration(L_i)$. The same argument holds for ν_{i3} as well. Also, $maxfrac(\nu_i) \leq 1 - s_{i-1,3} + f_{i,1,k}$ and hence $1 - \max \nu_i \geq s_{i-1,3} - f_{i,1,k} < e_{i,3} - s_{i,3} < duration(L_i)$. Therefore, by Proposition 1.2., L_i is executable from both ν_{i2} and ν_{i3}.

Notice that the so-constructed word w is not in L because all $b_{i,j}$ are disjoint. The word w' will be constructed almost the same way, with the only exception that the first repetition of the cycle is move not to $b_{i,2}$ but always the same interval, $b_{1,2}$. Its easy to see that Lemma 1 can be modified where $b_{i,2}$ is replaced everywhere by $b_{1,2}$. In particular this means that w contains an event at time $n + s_{1,2}$ for any $n \in \mathbb{N}$, and thus must be contained in L. Therefore, D has an accepting run on w' but the run on w' is visits the same sequence of states as the run of D on w. Therefore, D must accept w as well, which is a contradiction proving that L is not accepted by any deterministic Timed Automaton with Parity acceptance.

4   HD < ND

We now show that non-deterministic TA are more expressive than history-deterministic TA. In particular, we show that the following language L' over the singleton alphabet $\Sigma = \{a\}$ is recognised by a one-clock non-deterministic TA with reachability acceptance but not by any history-deterministic Parity TA. In words, L' asks to see two events a at unit distance. Formally,

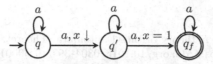

Fig. 3. A non-deterministic timed reachability automaton for L'.

$$L' \stackrel{\text{def}}{=} \{(\sigma_0, t_0)(\sigma_1, t_1)... \mid \exists i, j \in \mathbb{N}. \quad t_j - t_i = 1 \text{ and } \sigma_i = a \text{ and } \sigma_j = a\}.$$

The non-deterministic TA shown in Fig. 3 accepts the language L' by guessing positions i by reading an a, resetting a clock x and checking that it sees an a at distance 1.

Assume towards a contradiction that there exists a HD TA H with k clocks and maximum constant in guards c_x, that recognises L'. For all $i \leq k$ consider the finite word

$$w_i = \left(a, \frac{1}{k+1}\right) \cdots \left(a, \frac{k+1}{k+1}\right) \left(a, 1 + \frac{i}{k+1}\right)$$

that sees $k + 1$ equi-distant events in the interval $[0, 1]$ and then repeats the ith fractional value in the next integral interval. All these w_i are in L' and so the resolver gives a run on all such words. Note that the prefix up to time 1 is the same on all w_i and therefore the resolver gives the same run, on all of them until then. Consider the configuration ν reached by the resolver after reading the prefix up until and including the event $(a, 1)$. Since H has k clocks and $k + 1$ events a, there exists an $j \leq k$ such that $\nu(x) \neq 1 - \frac{j}{k+1}$ holds for all clocks x. That is, either no clock is reset while reading the jth event, or any clock reset at that time is again reset later. It follows that $\nu + \frac{j}{k+1} \sim \nu + \frac{j}{k+1} + \left(\frac{1}{2(k+1)}\right)$.

Finally, let's take the word

$$w' = \left(a, \frac{1}{k+1}\right) \left(a, \frac{2}{k+1}\right) \cdots \left(a, \frac{k+1}{k+1}\right) \left(a, 1 + \frac{j}{k+1} + \frac{1}{2(k+1)}\right)$$

Clearly w' is not in L'. However, H must have a run on w' which follows the accepting run of H on w_j. The final step in this run can be executed because the two runs end up in equivalent configurations. A contradiction.   □

We thus conclude that the classes of languages accepted by deterministic, history-deterministic and non-deterministic TAs are all different.

Funding. This work was supported in part by the ERC-2020-AdG 101020093, the EPSRC project EP/V025848/1, and the EPSRC project EP/X017796/1.

References

1. Alur, R., Dill, D.L.: A theory of timed automata. Theoret. Comput. Sci. **126**(2), 183–235 (1994). https://doi.org/10.1016/0304-3975(94)90010-8
2. Bagnol, M., Kuperberg, D.: Büchi good-for-games automata are efficiently recognizable. In: IARCS Annual Conference on Foundations of Software Technology and Theoretical Computer Science (FSTTCS). Leibniz International Proceedings in Informatics (LIPIcs), vol. 122, pp. 1–14. Schloss Dagstuhl-Leibniz-Zentrum fuer Informatik (2018). https://doi.org/10.4230/LIPIcs.FSTTCS.2018.16
3. Boker, U., Lehtinen, K.: Good for games automata: from nondeterminism to alternation. In: International Conference on Concurrency Theory (CONCUR). LIPIcs, vol. 140, pp. 1–16 (2019)
4. Boker, U., Lehtinen, K.: History determinism vs. good for gameness in quantitative automata. In: IARCS Annual Conference on Foundations of Software Technology and Theoretical Computer Science (FSTTCS). Leibniz International Proceedings in Informatics. (LIPIcs), vol. 213, pp. 1–20. Schloss Dagstuhl - Leibniz-Zentrum für Informatik (2021). https://doi.org/10.4230/LIPIcs.FSTTCS.2021.38
5. Boker, U., Lehtinen, K.: Token games and history-deterministic quantitative automata. In: FoSSaCS 2022. LNCS, vol. 13242, pp. 120–139. Springer, Cham (2022). https://doi.org/10.1007/978-3-030-99253-8_7
6. Colcombet, T.: The theory of stabilisation monoids and regular cost functions. In: International Colloquium on Automata, Languages and Programming (ICALP), pp. 139–150 (2009)
7. Guha, S., Jecker, I., Lehtinen, K., Zimmermann, M.: A bit of nondeterminism makes pushdown automata expressive and succinct. In: International Symposium on Mathematical Foundations of Computer Science (MFCS). Leibniz International Proceedings in Informatics (LIPIcs), vol. 202, pp. 1–20. Schloss Dagstuhl - Leibniz-Zentrum für Informatik (2021). https://doi.org/10.4230/LIPIcs.MFCS.2021.53
8. Henzinger, T.A., Kupferman, O., Rajamani, S.K.: Fair simulation. Inf. Comput. **173**(1), 64–81 (2002)
9. Henzinger, T.A., Lehtinen, K., Totzke, P.: History-deterministic timed automata. In: International Conference on Concurrency Theory (CONCUR) (2022)
10. Henzinger, T.A., Piterman, N.: Solving games without determinization. In: Ésik, Z. (ed.) CSL 2006. LNCS, vol. 4207, pp. 395–410. Springer, Heidelberg (2006). https://doi.org/10.1007/11874683_26
11. Herrmann, P.: Timed automata and recognizability. Inf. Process. Lett. **65**(6), 313–318 (1998). https://doi.org/10.1016/S0020-0190(97)00217-2
12. Kuperberg, D., Skrzypczak, M.: On determinisation of good-for-games automata. In: International Colloquium on Automata, Languages and Programming (ICALP), pp. 299–310 (2015)
13. Kupferman, O., Safra, S., Vardi, M.Y.: Relating word and tree automata. Ann. Pure Appl. Logic **138**(1–3), 126–146 (2006)
14. Lehtinen, K., Zimmermann, M.: Good-for-games ω-pushdown automata. Logical Methods in Computer Science 18 (2022)

Unambiguity and Fewness for Nonuniform Families of Polynomial-Size Nondeterministic Finite Automata

Tomoyuki Yamakami[✉]

Faculty of Engineering, University of Fukui, 3-9-1 Bunkyo, Fukui 910-8507, Japan
TomoyukiYamakami@gmail.com

Abstract. Nonuniform families of polynomial-size finite automata, which are series of indexed finite automata having polynomially many inner states, are used in the past literature to solve nonuniform families of promise decision problems. In such a nonuniform family, we focus our attention, in particular, on the variants of nondeterministic finite automata, which have at most "one" (unique or unambiguous), "polynomially many" (few) accepting computation paths, or unique/few computation paths leading to each fixed configuration. We prove that those variants of one-way machines are different in computational power. As for two-way machines restricted to instances of polynomially-bounded size, families of two-way polynomial-size nondeterministic finite automata are equivalent in power to families of unambiguous finite automata.

Keywords: Nonuniform state complexity · Finite automata · Accepting computation path · Unambiguous · Fewness

1 Background and an Overview

1.1 Historical Background: Unambiguity and Fewness in Complexity Theory

The number of accepting computation paths of an underlying nondeterministic machine has been a centerpiece of intensive research over the decades because the accepting criteria are of great importance for nondeterministic computation and it is indeed a key to the full understandings of nondeterministic computation.

The *unambiguity* of a language is in general characterized by an underlying machine, which has at most one accepting computation path. The study of unambiguous context-free languages, in particular, is also one of the important subjects in formal languages theory because there are efficient parsing algorithms for those languages.

In computational complexity theory, the unambiguity issues have been discussed since Valiant [15] introduced the unambiguous polynomial-time complexity class, known as UP, in connection to the existence of one-way functions.

© The Author(s), under exclusive license to Springer Nature Switzerland AG 2022
A. W. Lin et al. (Eds.): RP 2022, LNCS 13608, pp. 77–92, 2022.
https://doi.org/10.1007/978-3-031-19135-0_6

When we allow more than one accepting computation paths but limited to polynomially many paths, we then obtain FewP, which was introduced in [1,2]. It still remains open whether or not the inclusions $P \subseteq UP \subseteq FewP \subseteq NP$ are all proper. A series of papers in the past literature have proposed various refinements of unambiguous language families.

For the model of space-bounded machines, the logarithmic-space (or logspace) analogues of UP and FewP, denoted UL and FewL, were discussed in 1990 s s [5] for better understandings of the nondeterministic logarithmic-space computation. The space restriction sometimes presents a quite different landscape from the time restriction. With the help of the Karp-Lipton advice [10], for instance, Reinhardt and Allender [13] managed to prove the equivalence between NL/poly and UL/poly although NL and UL themselves are still unknown to coincide. Bourke, Tewari, and Vinodchandran [4] and lately also Pavan, Tewari, and Vinodchandran [12] gave an intriguing refinement of those complexity classes and they exhibited the existence of a rich structure among such refined complexity classes that are located between L and NL. Their refined classes associated with unambiguity and fewness notions include: ReachUL, ReachFewL, ReachLFew, and FewUL. It is also imperative to expand and explore the nature of unambiguity and fewness of accepting computation paths in other computational models.

1.2   Historical Background: Nonuniform Families of Finite Automata

Let us turn our attention to "finite(-state) automata", which are one of the simplest uniform models of computation. Those machines have been intensively studied since its early introduction but, only since late 1970s,s, nonuniform families of those machines have drawn our attention. In analogy to families of (Boolean) circuits, Sakoda and Sipser [14], following the work of Berman and Lingas [3], studied the families of polynomial-size finite automata, indexed by natural numbers. Later, a series of papers [6–9,17–20] have contributed to establishing a coherent theory over *nonuniform state complexity* of families of languages, more generally, promise decision problems. Nonuniform machine families are generally used as a vehicle to solve families of promise problems in quite efficient ways. In such a nonuniform setting, there are two parameters to take into consideration: machine's index n and input length $|x|$.

A family of finite automata is quite different from a family of (Boolean) circuits in the following key point: while each circuit in a circuit family takes only inputs of a fixed length, a finite automaton in an automata family can take inputs of arbitrary length. This makes it possible for us to discuss subfamilies of a family of automata by freely restricting the size of inputs, which is called a "ceiling". By choosing different ceilings, we can discuss the computational complexity of a wide variety of nonuniform families of finite automata. The notion of nonuniformity, ranging from advice-enhanced Turing machines to families of Boolean circuits, is as important as that of uniformity in computational complexity theory.

Families of promise problems that are solvable by nonuniform families of two-way polynomial-size deterministic finite automata are, in particular, denoted

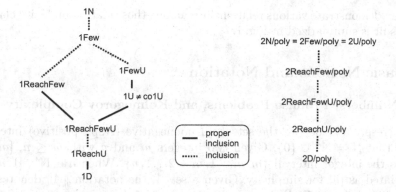

Fig. 1. Inclusion and collapse relations among families of promise problems discussed in this work.

collectively as 2D, and its nondeterministic variant is denoted 2N. In fact, the nonuniform nature of such machine families in fact provides enormous flexibility to solving families of promise problems. Manifestation of this fact has been demonstrated in the field of automata theory, for various machine types including deterministic, nondeterministic, probabilistic, quantum automata and also pushdown automata. The nonuniform families of finite automata have been studied in direct connection to logarithmic-space advised complexity classes, such as L/poly and NL/poly. It has been expected to further expand the scope of the research on nonuniform state complexity theory to other types of finite automata families.

1.3   New Challenges in This Work

Unfortunately, the theory of nonuniform state complexity has not yet intensively studied in depth and scope. Thus, it is imperative to replenish the theory by cultivating and examining structural properties of underlying finite automata. In this work, we intend to explore structural properties of machine models that lie between nonuniform families of deterministic finite automata and those of nondeterministic finite automata.

For this purpose, we need to adapt various notions related to unambiguity and fewness for nondeterministic logarithmic-space Turing machines [12] to fit into our setting of nonuniform families of finite automata. Our intension here is to replenish the theory by making new challenges in the topics of unambiguity and fewness for polynomial-size families of finite automata, because the notions of unambiguity and fewness are also important in automata theory. We intend to study the computational complexity of families of polynomial-size nondeterministic finite automata that satisfy various conditions on accepting computation concerning unambiguity and fewness.

In a spirit similar to [12], we will introduce six nonuniform state complexity classes between 1D and 1N in Sect. 3.1 and between 2D and 2N in Sect. 4. We will then demonstrate various relationships among those state complexity classes. Our result is summarized in Fig. 1.

2   Basic Notions and Notation

2.1   Numbers, Promise Problems, and Kolmogorov Complexity

Let $\mathbb{N}$ (resp., $\mathbb{N}^+$) denote the set of all nonnegative (resp., positive) integers. Notice that $\mathbb{N} = \mathbb{N}^+ \cup \{0\}$. Given two integers m and n with $m \leq n$, $[m, n]_{\mathbb{Z}}$ denotes the integer interval $\{m, m+1, m+2, \ldots, n\}$. When $n \in \mathbb{N}^+$, $[1, n]_{\mathbb{Z}}$ is abbreviated as $[n]$ for simplicity. Given a set A, the notation $\|A\|$ denotes the *cardinality* of A and $\mathcal{P}(A)$ denotes the *power set* of A.

A *polynomial* in this paper is assumed to have nonnegative integer coefficients. An *exponential* refers to a function of the form $2^{p(n)}$ for an appropriate polynomial p. Any *logarithm* is assumed to take the base 2 and any *logarithmic function* f has the form $a \log x + b$ for certain constants $a, b \geq 0$. For convenience, we say that a function f on $\mathbb{N}$ (i.e., from $\mathbb{N}$ to $\mathbb{N}$) *super-exponential* if, for any polynomial p, $f(n) > 2^{p(n)}$ holds for all but finitely many numbers $n \in \mathbb{N}$.

We briefly explain (nonuniform) families of promise (decision) problems as described in [17–20]. Given an alphabet Σ, a *promise (decision) problem* over Σ is a pair (A, B) of sets satisfying $A \cup B \subseteq \Sigma^*$ and $A \cap B = \varnothing$. Any instance in A is called *positive* and any instance in B is *negative*. To clarify the use of positive/negative instances, we often use the superscripts of $(+)$ and $(-)$ to express a promise problem as $(L^{(+)}, L^{(-)})$. We further consider a family $\mathcal{L}$ of promise problems $(L_n^{(+)}, L_n^{(-)})$ for all indices $n \in \mathbb{N}$; however, we focus only on families of promise problems over the "same" alphabet Σ. From this condition, we often denote the union $L_n^{(+)} \cup L_n^{(-)}$ by $\Sigma^{(n)}$, where the notation Σ^n is reserved for the set of strings of length n. Any string in $\Sigma^{(n)}$ indicates a *valid* (or a *promised*) instance over Σ. The *complement* of $\mathcal{L}$, denoted co-$\mathcal{L}$, consists of $(L_n^{(-)}, L_n^{(+)})$ for all indices $n \in \mathbb{N}$.

Let U denote any universal (deterministic) Turing machine taking binary input strings and eventually produces binary output strings. Given any binary strings x and y, the *conditional Kolmogorov complexity of x conditional to y*, denoted $C(x|y)$, is the length of the shortest binary string p such that U on inputs (p, y) produces x on its output tape. We use the notation ε to denote the *empty string*. If y is ε, we write $C(x)$ instead of $C(x|\varepsilon)$. This is referred to as the *(unconditional) Kolmogorov complexity of x*. See, e.g., [11] for more detail.

2.2   Families of Finite Automata

In this work, we use "standard" model of finite automata. We abbreviate a *one-way nondeterministic finite(-state) automaton* as a 1nfa and a *two-way nondeterministic finite(-state) automaton* as a 2nfa. Similarly, we call deterministic variants of 1nfa and 2nfa by 1dfa and 2dfa, respectively.

We consider "families" of finite automata of the same machine type over the same alphabet Σ. In particular, we use families $\mathcal{M} = \{M_n\}_{n\in\mathbb{N}}$ of 2nfa's (as well as 2dfa's, 1nfa's, and 1dfa's) as a base model to solve families of promise problems.

Each finite automaton M_n in the family $\mathcal{M}$ is expressed as a septuple $(Q_n, \Sigma, \{\triangleright, \triangleleft\}, \delta_n, q_{0,n}, Q_{acc,n}, Q_{rej,n})$ with two designated endmarkers $\triangleright$ and $\triangleleft$ and two sets $Q_{acc,n}$ and $Q_{rej,n}$ of accepting (inner) states and rejecting (inner) states satisfying both $Q_{acc,n} \cup Q_{rej,n} \subseteq Q_n$ and $Q_{acc,n} \cap Q_{rej,n} = \varnothing$. Any inner state in $Q_{acc,n} \cup Q_{rej,n}$ are simply called a *halting (inner) state*. The transition function δ is defined only on non-halting states. The *state complexity* of M_n refers to $|Q_n|$, which is the total number of M_n's inner states, and it is denoted $sc(M_n)$. A family $\mathcal{M}$ is said to have *polynomial size* if there exists a polynomial p such that $sc(M_n) \leq p(n)$ holds for all $n \in \mathbb{N}$. For more information on the underlying setting of automata families, the reader refers to [18].

Following [18], a one-way finite automaton is always assumed to make no stationary move (or ε-move); that is, its tape head must move only in one direction, to the right, whenever it reads an input symbol (including the endmarkers). This condition is sometimes called *real time* in the literature.

It is imperative to clarify a few important terminologies associated with "computation" of finite automata M_n on input x. A *surface configuration* of M_n on x is of the form (q, i) in $Q \times [0, |x| + 1]_{\mathbb{Z}}$ excluding x, which indicates that M_n is in inner state q and its tape head is located at cell i, assuming that tape cells are indexed by nonnegative integers and that $\triangleright$ and $\triangleleft$ are placed respectively at cell 0 and cell $|x| + 1$. In the rest of this work, since we deal only with surface configurations, we drop the word "surface" altogether from "surface configurations". With the use of configurations, we consider a *computation graph* of M_n on x, whose vertices are configurations of M_n on x and a transition from any non-halting configuration to another configuration forms a directed edge. We write $(p, i) \vdash (q, i + 1)$ to express an edge of this graph. A *computation path* is a path in a computation graph from the root (i.e., the initial configuration) to a certain leaf (i.e., a halting configuration) if any. A computation path is *accepting* (resp., *rejecting*) if it ends in a configuration with an accepting (resp., a rejecting) configuration.

We say that M_n *halts* on x if there exists a computation path of finite length in the computation graph of M_n on x. In this work, we are interested only in finite automata that always halt on all valid instances. For such a halting 2nfa M_n, we say that M_n *accepts* x if M_n starts with $\triangleright x \triangleleft$ and produces a certain accepting computation path, and M is said to *reject* x otherwise.

Let us consider a family $\mathcal{L} = \{(L_n^{(+)}, L_n^{(-)})\}_{n\in\mathbb{N}}$ of promise problems over a fixed alphabet Σ. Recall that $\Sigma^{(n)}$ expresses the set of all valid (or promised) instances over Σ for $(L_n^{(+)}, L_n^{(-)})$. For any other "invalid" instance x, when a machine, say, M_n takes x as an input, we do not require any condition on the behavior of the machine. A family $\mathcal{M} = \{M_n\}_{n\in\mathbb{N}}$ of machines over Σ is said to *solve* (or *recognize*) $\mathcal{L}$ if, for any $n \in \mathbb{N}$, (1) for any $x \in L_n^{(+)}$, M_n accepts x and

(2) for any $x \in L_n^{(-)}$, M_n rejects x. There may be a case where M does not even halt on invalid instances.

The notation 2N (resp., 1N) denotes the collection of all families of promise problems solved by appropriate families of polynomial-size 2nfa's (resp., polynomial-size 1nfa's). Similarly, the notation 2D (resp., 1D) is defined using 2dfa's (resp., 1dfa's).

2.3   Unambiguous and Few Computation Paths

Unambiguous and few computation paths of nondeterministic machines have played an important role in computational complexity theory. Following early works of [2,5,12,15], we introduce key notions related to "unambiguity" and "fewness" into theory of nonuniform state complexity. Let us recall that each 1nfa is allowed to have multiple accepting states and multiple rejecting states. Let $\mathcal{M} = \{M_n\}_{n \in \mathbb{N}}$ denote any family of nondeterministic finite automata.

o Firstly, $\mathcal{M}$ is called *unambiguous* if, for each index $n \in \mathbb{N}$ and for any input $x \in \Sigma^{(n)}$, there is at most one accepting computation path of M_n on x.
  Similarly, $\mathcal{M}$ is *weak-unambiguous* if, for any index $n \in \mathbb{N}$, for any input $x \in \Sigma^{(n)}$, and for any accepting configuration $conf$, there exists at most one computation path from the initial configuration of M_n on x to $conf$.
o Moreover, $\mathcal{M}$ is *reach-unambiguous* if, for any index $n \in \mathbb{N}$, for any input $x \in \Sigma^{(n)}$, and for any configuration $conf$, there is at most one computation path from the initial configuration of M_n on x to $conf$.
o In contrast, $\mathcal{M}$ is *accept-few*[1] if there exist a polynomial p such that, for any index $n \in \mathbb{N}$ and for any input $x \in \Sigma^{(n)}$, there are at most $p(n, |x|)$ accepting computation paths of M_n on x.
o Lastly, $\mathcal{M}$ is *reach-few* if there exist a polynomial p such that, for any index $n \in \mathbb{N}$, for any input $x \in \Sigma^{(n)}$, and for any configuration $conf$ of M_n on x, there are at most $p(n, |x|)$ computation paths of M_n on x from the initial configuration to $conf$.

We remark that all the above five conditions are applied only to "valid" inputs and there is no requirement for "invalid" inputs.

3   Complexity Classes Defined by One-Way Head Moves

We begin with discussing the computational complexity of families of promise problems solved by nonuniform families of one-way nondeterministic finite automata of various types introduced in Sect. 2.3.

It is important to remark that, since tape heads of 1nfa's always move to the right without making stationary moves, we can modify the 1nfa's so that, whenever they fail to enter accepting states until reading the right endmarker

---

[1] This notion is called just "few" in [12]. For clarity reason, here we use a slightly different term.

◁, they must enter rejecting states at the time of reading ◁. Moreover, when the 1nfa's enter accepting states even before reading ◁, it is also possible to postpone the timing of acceptance until reading ◁. The 1nfa's obtained by these modifications halt precisely at reading ◁.

3.1 Definitions of New Complexity Classes

In a way similar to [2,5,12,15], as various subfamilies of 1N, we introduce six nonuniform state complexity classes associated with unambiguity and fewness of accepting computation paths given in Sect. 2.3.

○ 1Few consists of all families of promise problems, each family of which is solved by an appropriate family of 1-way *accept-few* finite automata having polynomially many inner states.
○ 1ReachFew is a unique subclass of 1Few, whose underlying finite automata are additionally *reach-few*.
○ 1ReachFewU is a unique subclass of 1ReachFew, whose underlying finite automata are additionally *unambiguous*.
○ 1ReachU is a unique subclass of 1ReachFewU, whose underlying finite automata are additionally *reach-unambiguous*.
○ 1FewU is defined from 1Few with underlying finite automata are additionally *weak-unambiguous*.
○ 1U consists of all families of promise problems, each family of which is solved by an appropriate family of 1-way *unambiguous* finite automata having polynomially many inner states.

The following inclusion relationships hold among the above-mentioned complexity classes. See also Fig. 1.

Lemma 1. *(1)* 1D ⊆ 1ReachU ⊆ 1ReachFewU ⊆ 1ReachFew ⊆ 1Few. *(2)* 1ReachFewU ⊆ 1U ⊆ 1FewU ⊆ 1Few ⊆ 1N.

3.2 Class Separations

In what follows, we will demonstrate the class separations depicted in Fig. 1. For our later argument, we will introduce several notions and notation.

Given numbers $i_1, i_2, \ldots, i_k$ in $\mathbb{N}^+$, $[i_1, i_2, \ldots, i_k]$ denotes the binary string $1^{i_1} 0 1^{i_2} 0 \cdots 0 1^{i_k}$. The value k is called the *size* of $[i_1, i_2, \ldots, i_k]$. For any string r of the form $[i_1, i_2, \ldots, i_k]$, it follows that $|r| = \sum_{j=1}^{k} i_j + k - 1$. Given such a string r and for any index $e \in [k]$, the notation $(r)_{(e)}$ denotes i_e and $Set(r)$ stands for the set $\{i_1, i_2, \ldots, i_k\}$. Let A_n denote the set of all strings of the form $[i_1, i_2, \ldots, i_k]$ for certain numbers $k \in \mathbb{N}^+$ and $i_1, i_2, \ldots, i_k \in [n]$. Additionally, we set $A_n(k) = \{r \in A_n \mid \text{ size of } r \text{ is } k \}$ for each fixed number $k \in \mathbb{N}^+$.

As the first class separation, we show that 1ReachU properly contains 1D.

Lemma 2. $1D \neq 1ReachU$.

Proof. Given each index $n \in \mathbb{N}$, we set $L_n^{(+)} = \{r_1 \# r_2 \mid r_1, r_2 \in A_n(n), \exists! e \in [n][(r_1)_{(e)} \neq (r_2)_{(e)}]\}$ and $L_n^{(-)} = \{r_1 \# r_2 \mid r_1, r_2 \in A_n(n), \forall e \in [n][(r_1)_{(e)} = (r_2)_{(e)}]\}$. We then denote by $\mathcal{L}_1$ the family $\{(L_n^{(+)}, L_n^{(-)})\}_{n \in \mathbb{N}}$.

To show that $\mathcal{L}_1$ is in 1ReachU, let us consider the following 1nfa N_n for each $n \in \mathbb{N}$: on input of the form $r_1 \# r_2$ with $r_1, r_2 \in A_n(n)$, guess (i.e., nondeterministically choose) an index $e \in [n]$, read through r_1 to find the eth entry $i_e = (r_1)_{(e)}$, remember it as an inner state of the form (i_e, e), move to r_2, and check whether $i_e = (r_2)_{(e)}$. If so, accept the input, or else reject it. It is not difficult to see that N_n solves $(L_n^{(+)}, L_n^{(-)})$. By the definition of $L_n^{(+)}$, N_n is unambiguous on all valid instances. Since N_n is also reach-unambiguous, $\mathcal{L}_1$ belongs to 1ReachU.

Next, we want to show that $\mathcal{L}_1 \notin 1D$ by way of contradiction. Assume that $\mathcal{L}_1 \in 1D$. This implies that co-$\mathcal{L}_1 \in 1D$ since $1D = co\text{-}1D$. Take a family $\mathcal{M} = \{M_n\}_{n \in \mathbb{N}}$ of 1dfa's having polynomially many inner states that solves co-$\mathcal{L}_1$. Since M_n solves $(L_n^{(-)}, L_n^{(+)})$ for each $n \in \mathbb{N}$, M_n accepts $r_1 \# r_2$ in $L_n^{(-)}$ and M_n rejects $r_1 \# r_2$ in $L_n^{(+)}$. Since $\mathcal{M}$ has polynomial size, there exists a constant $c > 0$ satisfying $|Q_n| \leq n^c$ for any index $n \in \mathbb{N}$. Here, we take a sufficiently large number $n \in \mathbb{N}$ for which $n > c$ holds.

For each string $r_1 \in A_n(n)$, $\mu(r_1)$ denotes the unique inner state q of M_n obtained just after reading $r_1 \#$. It follows that, for any pair $r_1, r_1' \in A_n(n)$, $\mu(r_1) = \mu(r_1')$ implies $r_1 = r_1'$. This is shown as follows. Assume that $\mu(r_1) = \mu(r_1')$ and $r_1 \neq r_1'$. Take an index $e \in [n]$ satisfying $(r_1)_{(e)} \neq (r_1')_{(e)}$. Consider two strings $r_1 \# r_1$ and $r_1' \# r_1$. Since $r_1 \# r_1 \in L_n^{(-)}$ and M_n is deterministic, $r_1' \# r_1$ must be accepted. Thus, we obtain $r_1' \# r_1 \in L_n^{(-)}$, a contradiction. Therefore, we conclude that $|A_n(n)| \leq |Q_n|$. However, since $|A_n(n)| = n^n$ and $|Q_n| \leq n^c$, it follows that $n \leq c$. This is a clear contradiction.   □

Next, we look into a relationship between 1U and 1FewU.

Lemma 3. $1U \neq 1FewU$.

Proof. We first define a family $\mathcal{L}_2 = \{(L_n^{(+)}, L_n^{(-)})\}_{n \in \mathbb{N}}$ by setting $L_n^{(+)} = \{r_1 \# r_2 \mid r_1, r_2 \in A_n(n), \exists e \in [n][(r_1)_{(e)} \neq (r_2)_{(e)}]\}$ and $L_n^{(-)} = \{r_1 \# r_2 \mid r_1, r_2 \in A_n(n), \forall e \in [n][(r_1)_{(e)} = (r_2)_{(e)}]\}$.

Let us prove that $\mathcal{L}_2$ is in 1FewU. This can be shown by the following 1nfa N_n. We assume that this machine has n accepting states $\hat{q}_1, \hat{q}_2, \ldots, \hat{q}_n$. On input $x = r_1 \# r_2$ with $r_1, r_2 \in A_n(n)$, guess an index $e \in [n]$, read r_1, remember the eth number $i_e = (r_1)_{(e)}$ by entering inner states of the form (i_e, e), read r_2, and check whether $i_e = (r_2)_{(e)}$. If so, enter the eth accepting state $\hat{q}_e$. It follows that, for each accepting state $\hat{q}_e$, there are at most one accepting computation path leading to $\hat{q}_e$. Thus, N_n is accept-few and also weak-unambiguous. Since N_n solves $(L_n^{(+)}, L_n^{(-)})$ for all $n \in \mathbb{N}$, $\mathcal{L}_2$ belongs to 1FewU.

Next, we want to show that $\mathcal{L}_2 \notin 1U$. Toward a contradiction, we take a family $\mathcal{M} = \{M_n\}_{n \in \mathbb{N}}$ of polynomial-size unambiguous 1nfa's and assume that

$\mathcal{M}$ solves $\mathcal{L}_2$. Since M_n is unambiguous, it is possible to assume that $|Q_{acc,n}| = 1$ for all $n \in \mathbb{N}$.

Assume that $|Q_n| \leq n^c$ for a certain constant $c \geq 1$. Choose any large enough number $n \in \mathbb{N}$ satisfying $n^c < 2^n$. In what follows, we consider only inputs of the form $r_1 \# r_2$ in $L_n^{(+)}$ with $r_1, r_2 \in A_n(n)$. Recall the notation $C(x)$ from Sect. 2.1. Since $r_1 \in A_n(n)$, r_1 has the form $[i_1, i_2, \ldots, i_n]$ for $i_1, i_2, \ldots, i_n \in [n]$. Letting $I_n = \{i \in [n] \mid i \geq \sqrt{n}\}$, if $i_1, i_2, \ldots, i_n$ are all taken from I_n, then it follows that $|r_1| - n = \sum_{j=1}^{n} i_j - 1 \geq n\sqrt{n} - 1$. Hence, we obtain $\log(|r_1| - n) \geq \log n$. Take a sufficiently large number $n \in \mathbb{N}$ and fix a string $r_1 \in A_n(n)$ for which $C(r_1) \geq \log^2(|r_1| - n) \geq \log^2 n$ holds.

We partition r_1 into $r_{11} r_{12}$ for which $|r_{12}| > 2c \log n$ holds. Note that $r_1 \# r_{11} r_{12} \notin L_n^{(+)}$. Note that M_n is unambiguous on all valid instances. We define $Q_{r_1, r_{11}}$ to be the collection of all inner states $q \in Q_n$ such that there exists a string w_{12} with $w_{12} \neq r_{12}$ for which $r_1 \# r_{11} w_{12} \in L_n^{(+)}$ and, along a unique accepting computation path γ_1 on $r_1 \# r_{11} w_{12}$, M_n enters q just after reading $r_1 \# r_{11}$. Similarly, we set $\bar{Q}_{r_1, r_{12}}$ to be the set of all inner states $q \in Q_n$ such that there exists a string w_{11} with $w_{11} \neq r_{11}$ for which $r_1 \# w_{11} r_{12} \in L_n^{(+)}$ and, along a unique accepting computation path γ_2 on $r_1 \# w_{11} r_{12}$, M_n enters q just after reading $r_1 \# w_{11}$. We then claim the following.

Claim 4 $Q_{r_1, r_{11}} \cap \bar{Q}_{r_1, r_{12}} = \varnothing$.

Proof. Assume otherwise and take an inner state $q \in Q_{r_1, r_{11}} \cap \bar{Q}_{r_1, r_{12}}$. By the definition of $Q_{r_1, r_{12}}$, there is a string $w_{12} \neq r_{12}$ and a unique accepting computation path γ_1 of M_n on $r_1 \# r_{11} w_{12}$, on which M_n enters q after reading $r_1 \# r_{11}$. Similarly, there is a string $w_{11} \neq r_{11}$ and a unique accepting computation path γ_2 of M_n on $r_1 \# w_{11} r_{12}$, on which M_n enters q after reading $r_1 \# w_{11}$. Consider the following computation of M_n. We first follow γ_1 until M_n reads off $r_1 \# r_{11}$ and enters q. Starting with q, we read off r_{12} by following γ_2. Since γ_2 is an accepting computation path, we eventually enter a certain accepting state. This implies that we accept $r_1 \# r_{11} r_{12}$, and thus it belongs to $L_n^{(+)}$, a contradiction. □

For convenience, let $l_1 = |r_{11}|$ and $l_2 = |r_{12}|$. Since $Q_{r_1, r_{11}} \cap \bar{Q}_{r_1, r_{12}} = \varnothing$, take two accepting computation paths γ_1 and γ_2 and two inner states $p_1 \in Q_{r_1, r_{11}}$ and $p_2 \in \bar{Q}_{r_1, r_{12}}$ appearing on them, respectively. Let q_{acc} denote a unique accepting state that M_n enters after reading $r_1 \# r_{11} r_{12}$. Moreover, let q denote an inner state that M_n enters just after reading off $r_1 \#$.

Hereafter, we intend to construct r_1 from certain information of size significantly less than $\log^2 n$. Starting with q, we can find r_{11} uniquely if we know (n, l_1, l_2, p_1). This can be done by cycling through all strings z of length l_1 and running M_n on z to reach p_1 since there is a unique accepting computation path, on which M_n reaches p_1 after reading off r_{11}. Similarly, from p_2, we can find r_{12} since an accepting computation path uniquely determines r_{12}. Thus, for the construction of $r_1 = r_{11} r_{12}$, it suffices to know the tuple $(n, q_0, q, p_1, p_2, q_{acc}, l_1, l_2)$. Therefore, we obtain $C(r_1) \leq |q_0| + |p| + |p_1| + |p_2| + |q_{acc}| + O(\log n) \leq O(\log n)$, a contradiction against the bound of $C(r_1) \geq \log^2 n$. □

The next lemma easily follows from the proof of Lemma 3.

Lemma 5. 1ReachFew $\not\subseteq$ 1U.

Proof. Let us recall the family $\mathcal{L}_2$ of promise problems defined in the proof of Lemma 3. Recall also the family of 1nfa's used in the same proof to show the membership "$\mathcal{L}_2 \in$ 1FewU". These 1nfa's are accept-few and unambiguous. Moreover, they are also reach-few. Thus, $\mathcal{L}_2$ belongs to 1ReachFew. Note that, by the proof of Lemma 3, $\mathcal{L}_2 \notin$ 1U. □

Lemma 5 contrasts the result of Pavan et al. [12], who demonstrated that ReachFewL is included in UL $\cap$ co-UL.

Corollary 6. 1ReachFewU $\neq$ 1ReachFew

Proof. The proof of Lemma 5 asserts that $\mathcal{L}_2 \in$ 1ReachFew. Recall from the proof of Lemma 3 that $\mathcal{L}_2 \notin$ 1U. Because of 1ReachFewU $\subseteq$ 1U, we conclude that $\mathcal{L}_2 \notin$ 1ReachFewU. □

In addition to the class separations given by Lemmas 2–3 and Corollary 6, we show the non-closure property of 1U.

Lemma 7. *The family* 1U *is not closed under complementation.*

Proof. We recall the family $\mathcal{L}_1 = \{(L_n^{(+)}, L_n^{(-)})\}_{n \in \mathbb{N}}$ of promise problems from the proof of Lemma 2, in which $\mathcal{L}_1$ is shown to be in 1ReachU. Since 1ReachU $\subseteq$ 1U, we instantly obtain $\mathcal{L}_1 \in$ 1U.

Recall that co-$\mathcal{L}_1$ is of the form $\{(L_n^{(-)}, L_n^{(+)})\}_{n \in \mathbb{N}}$. Hereafter, we show that co-$\mathcal{L}_1 \notin$ 1U. To lead to a contradiction, let us assume the existence of a nonuniform family $\mathcal{M} = \{M_n\}_{n \in \mathbb{N}}$ of polynomial-size unambiguous 1nfa's solving co-$\mathcal{L}_1$. Recall that $L_n^{(+)} = \{r_1 \# r_2 \mid r_1, r_2 \in A_n(n), \exists! e \in [n][(r_1)_{(e)} \neq (r_2)_{(e)}]\}$ and $L_n^{(-)} = \{r_1 \# r_2 \mid r_1, r_2 \in A_n(n), \forall e \in [n][(r_1)_{(e)} = (r_2)_{(e)}]\}$.

The following argument is similar in essence to the proof of Lemma 2. Choose any sufficiently large n so that $n^c < 2^n$ holds. Take $(L_n^{(-)}, L_n^{(+)})$ and M_n, and focus on all inputs x of the form $r_1 \# r_2$ in $\Sigma^{(n)}$ $(= L_n^{(+)} \cup L_n^{(-)})$. Note that, for any pair $r_1, r_2 \in A_n(n)$, if M_n accepts $r_1 \# r_2$, then there exists a unique $q \in Q_n$ such that M_n enters q just after reading $r_1 \#$ and then enters an accepting state after reading r_2. We write $\mu(r_1)$ for this inner state q. Note that $|Q_n| < |A_n(n)|$ since $|A_n(n)| = n^n$. From $|Q_n| < |A_n(n)|$, it follows that there exist two elements $r_1, r_1' \in A_n(n)$ with $r_1 \neq r_1'$ satisfying $\mu(r_1) = \mu(r_1')$. This implies that $r_1 \# r_1'$ is accepted by M_n, and thus it belongs to $L_n^{(-)}$. This is a clear contradiction. □

For a class $\mathcal{C}$ of families of promise problems, we write co-$\mathcal{C}$ for the complexity class $\{\mathcal{L} \mid$ co-$\mathcal{L} \in \mathcal{C}\}$. It is known that co-1D coincides with 1D. Lemma 7 then asserts that 1U $\neq$ co-1U.

4 Complexity Classes Defined by Two-Way Head Moves

In Sect. 3.1, we have introduced six nonuniform state complexity classes situated in between 1D and 1N. By replacing underlying one-way finite automata defining those complexity classes with two-way finite automata, we naturally obtain the corresponding six nonuniform state complexity classes: 2U, 2ReachU, 2ReachFewU, 2Few, 2FewU, and 2ReachFew.

4.1 Case of Logarithmic and Polynomial Ceilings

In comparison to Sect. 3.1, we focus our study on the case of two-way head moves of 1npda's.

A family $\mathcal{L} = \{(L_n^{(+)}, L_n^{(-)})\}_{n \in \mathbb{N}}$ of promise problems is said to *have a polynomial ceiling* if there exists a polynomial p for which $\Sigma^{(n)} \subseteq \Sigma^{\leq p(n)}$ holds for all $n \in \mathbb{N}$. Similarly, we can define the notion of *exponential ceiling* (resp., *logarithmic ceiling*) simply by replacing the term "polynomial" with "exponential" (resp., "logarithmic function"). More generally, given a function $g : \mathbb{N} \to \mathbb{N}$, we say that $\mathcal{L}$ has an $g(n)$-ceiling if $\Sigma^{(n)} \subseteq \Sigma^{\leq g(n)}$ follows for all $n \in \mathbb{N}$.

For two functions f, g on $\mathbb{N}$ (i.e., from $\mathbb{N}$ to $\mathbb{N}$), we say that g *majorizes* f (denoted $g \geq f$) if $g(n) \geq f(n)$ holds for all $n \in \mathbb{N}$. Consider a family $\mathcal{L} = \{(L_n^{(+)}, L_n^{(-)})\}_{n \in \mathbb{N}}$ of promise problems having an $f(n)$-ceiling. It then follows that $L_n^{(+)} = \{x \in L_n^{(+)} \mid |x| \leq f(n)\}$ and $L_n^{(-)} = \{x \in L_n^{(-)} \mid |x| \leq f(n)\}$. For any function g on $\mathbb{N}$ that majorizes f, since $\Sigma^{(n)} \subseteq \Sigma^{\leq f(n)} \subseteq \Sigma^{\leq g(n)}$, $\mathcal{L}$ has a $g(n)$-ceiling as well.

Given a function f on $\mathbb{N}$, the notation $2\mathrm{N}/f(n)$ stands for the subclass of $2\mathrm{N}$, consisting of all families of promise problems having $f(n)$-ceilings. Notice that $g \geq f$ implies $2\mathrm{N}/f(n) \subseteq 2\mathrm{N}/g(n)$. For a set $\mathcal{F}$ of functions on $\mathbb{N}$, $2\mathrm{N}/\mathcal{F}$ denotes the union of all $2\mathrm{N}/f(n)$ for any $f \in \mathcal{F}$. In a similar way, we obtain $2\mathrm{D}/\mathcal{F}$ from $2\mathrm{D}$. These terminologies are also applied to 2FewU, 2ReachU, 2ReachFew, and 2ReachFewU.

For convenience, we abbreviate the sets of logarithmic functions, polynomials, exponentials, and sub-exponentials, as "log", "poly", "exp", and "subexp", respectively. This subsection will concentrate on the case where $\mathcal{F}$ is one of those sets of functions. We start with an easy case of $\mathcal{F} = \log$.

Lemma 8. $2\mathrm{N}/\log = 2\mathrm{D}/\log$. *More strongly,* $2\mathrm{N}/\log \subseteq 1\mathrm{D}$.

Proof. We show that $2\mathrm{N}/\log \subseteq 1\mathrm{D}$ because this implies $2\mathrm{N}/\log \subseteq 1\mathrm{D}/\log \subseteq 2\mathrm{D}/\log$. Let $\mathcal{L} = \{(L_n^{(+)}, L_n^{(-)})\}_{n \in \mathbb{N}}$ be any family of promise problems in $2\mathrm{N}/\log$. Take a function $\ell(x)$ of the form $a \log x + b$ for two constants $a, b \geq 0$. Notice that $\Sigma^{(n)} \subseteq \Sigma^{\leq \ell(n)}$, where $\Sigma^{(n)} = L_n^{(+)} \cup L_n^{(-)}$, and $|\Sigma^{\ell(n)}| = O(n^a)$. It then follows that $|\Sigma^{\leq \ell(n)}| = n^{O(1)}$.

Fix $n \in \mathbb{N}$. We enumerate all elements in $\Sigma^{\leq \ell(n)}$ as $x_0^{(n)}, x_1^{(n)}, x_2^{(n)}, \ldots$ according to the lexicographic order. We define Q_n to be composed of all strings of the form $[\begin{smallmatrix} x_i^{(n)} \\ a_i \end{smallmatrix}]$, where (i) $a_i = +1$ if $x_i^{(n)} \in L_n^{(+)}$, (ii) $a_i = -1$ if $x_i^{(n)} \in L_n^{(-)}$, and (iii) $a_i = 0$ otherwise. Let us design a 1dfa M_n to read an entire input, say, x, determine i satisfying $x = x_i^{(n)}$, and accept (resp., reject) x if $a_i = +1$ (resp., $a_i = -1$). The family $\mathcal{M} = \{M_n\}_{n \in \mathbb{N}}$ clearly has polynomial size. Since $\mathcal{M}$ solves $\mathcal{L}$, $\mathcal{L}$ belongs to 1D. □

We want to prove the following collapse result. Our proof is motivated by a simulation technique of [13].

Theorem 9. 2U/poly = 2FewU/poly = 2Few/poly = 2N/poly.

However, it is not known whether 2N/subexp = 2U/subexp, 2N/exp = 2U/exp, or even 2N = 2U.

For two other nonuniform state complexity classes, 2ReachFewU and 2ReachFew, we obtain the following relationships given in Fig. 1.

Corollary 10. 2D/poly $\subseteq$ 2ReachU/poly $\subseteq$ 2ReachFewU/poly $\subseteq$ 2ReachFew/poly $\subseteq$ 2U/poly.

Proof of Theorem 9. Because 2U/poly $\subseteq$ 2FewU/poly $\subseteq$ 2Few/poly $\subseteq$ 2N/poly, it suffices to prove the inclusion 2N/poly $\subseteq$ 2U/poly, which is equivalent to 2N/poly $\subseteq$ 2U. We hereafter prove that 2N/poly $\subseteq$ 2U.

It is shown in [8] (re-proven in [18]) that 2N/poly $\subseteq$ 2D iff NL $\subseteq$ L/poly. In a similar vein, we claim the following.

Claim 11. NL $\subseteq$ UL/poly *implies* 2N/poly $\subseteq$ 2U.

Note that Reinhardt and Allender [13] proved the inclusion NL $\subseteq$ UL/poly. Assuming that Claim 11 is true, we instantly obtain 2N/poly $\subseteq$ 2U from their result. Therefore, in what follows, we aim at proving Claim 11. For this purpose, we adopt the proof technique of [17–20] and introduce the notion of parameterized decision problems.

A *parameterized decision problem* is a pair (L, m) of a language L over alphabet Σ and a size parameter m mapping Σ^* to $\mathbb{N}$. A *log-space size parameter* m must satisfy the condition that there exists a log-space deterministic Turing machine (DTM) M for which M on input x produces $1^{m(x)}$ on a write-once output tape. An advice function h maps $\mathbb{N}$ to $\mathbb{N}$ and it is said to be *polynomially bounded* if $h(n) \leq p(n)$ holds for all $n \in \mathbb{N}$ for an appropriately chosen polynomial p. A parameterized decision problem (L, m) belongs to PHSP if m is *polynomially honest* (that is, there exists a polynomial q such that $|x| \leq q(m(x))$ for all strings x). The notation para-NL/poly denotes the class composed of all parameterized decision problems (L, m) with log-space size parameters m, such that, for each (L, m), there exists a nondeterministic Turing machine (NTM) M equipped with read-only input and advice tapes and a polynomially-bounded advice function h for which M solves L in time $(|x|m(x))^{O(1)}$ using space $O(\log m(x))$ with access to advice strings $h(|x|)$ written on the advice tape, where x is a "symbolic" input. Similarly, para-UL/poly is defined using unambiguous NTMs (or UTM, for short) instead of "standard" NTMs. The reader refers to [16–18] for more information on parameterized problems.

Claim 11 follows immediately from the following two assertions. Given a language L over alphabet Σ, its complement $\Sigma^* - L$ is succinctly denoted $\overline{L}$.

Claim 12. *1.* NL $\subseteq$ UP/poly *implies para-*NL/poly $\cap$ PHSP $\subseteq$ *para-*UL/poly.
*2. para-*NL/poly $\cap$ PHSP $\subseteq$ *para-*UL/poly *implies* 2N/poly $\subseteq$ 2U.

Proof Sketch. We loosely follow an argument made in the proof of [18, Proposition 5.1].

(1) Assume that NL $\subseteq$ UL/poly. Note that NL $\subseteq$ UP/poly implies NL/poly $\subseteq$ UL/poly. Let us consider an arbitrary parameterized decision problem (L, m) in para-NL/poly$\cap$PHSP with a log-space size parameter $m : \Sigma^* \to \mathbb{N}$ and a language L over alphabet Σ. Take a polynomial-size advice function h and an NTM M_0 such that M_0 recognizes $\{(x, h(|x|)) \mid x \in L\}$ in time $(|x|m(x))^{O(1)}$ and space $O(\log m(x))$. We wish to verify that $(L, m) \in$ para-UL/poly. For this purpose, we define $L_n^{(+)} = L \cap \Sigma_n$ and $L_n^{(-)} = \overline{L} \cap \Sigma_n$, where $\Sigma_n = \{x \in \Sigma^* \mid m(x) = n\}$. We further set $\mathcal{L} = \{(L_n^{(+)}, L_n^{(-)})\}_{n \in \mathbb{N}}$.
Since m is polynomially honest, we take a polynomial q such that $|x| \leq q(m(x))$ for all x. We write K' for the set $\{(x, 1^t) \mid x \in L, t \in \mathbb{N}, m(x) \leq t\}$ and claim that $K' \in$ NL/poly. Consider the following algorithm. On input $(x, 1^t)$, we check if $m(x) \leq t$ using log space. This is possible because m is log-space computable. If $m(x) > t$, then we reject the input; otherwise, we simulate M_0 on $(x, h(|x|))$. Clearly, this algorithm recognizes K'. This algorithm can be realized by an appropriate NTM running in time $(|x|t)^{O(1)}$ and space $O(\log |x|t)$ with the help of h. Since $|(x, 1^t)| = O(|x| + t)$, K' belongs to NL/poly.
Since NL/poly $\subseteq$ UL/poly by our assumption, K' must be in UL/poly. Take a UTM M, an advice function g, a logarithmic function ℓ' such that M recognizes K' using at most $\ell'(|z|)$ space with an access to $g(|z|)$, where z indicates a "symbolic" input. We then design a new NTM N for (K, m') as follows. Define $g'(|x|) = g(|(x, 1^t)|)$ for all $x \in \Sigma^*$. On input x, compute $n = m(x)$, and run M on $(x, 1^n)$ with $g'(|x|)$. Note that N's space usage is $O(\ell'(|x| + n) + \log |x|) \subseteq O(\log(q(m(x)) + n) + \log |x|) \subseteq O(\log m(x))$ since q is a polynomial satisfying $|x| \leq q(m(x))$. This shows that $(L, m) \in$ para-UL/poly.

(2) Assume that para-NL/poly $\cap$ PHSP $\subseteq$ para-UL/poly. Let $\mathcal{L} = \{(L_n^{(+)}, L_n^{(-)})\}_{n \in \mathbb{N}}$ be an arbitrary element of 2N/poly and take two polynomials p, q and a family $\mathcal{M} = \{M_n\}_{n \in \mathbb{N}}$ of 2nfa's that solves $\mathcal{L}$ with the following properties: each M_n has at most $p(n)$ inner states and $\Sigma^{(n)}$ $(= L_n^{(+)} \cup L_n^{(-)})$ is included in $\Sigma^{\leq q(n)}$. Hereafter, we show that $\mathcal{L}$ belongs to 2U.
For each index $n \in \mathbb{N}$, we define $K_n^{(+)} = \{1^n \# x \mid x \in L_n^{(+)}\}$ and $K_n^{(-)} = \{1^n \# x \mid x \in L_n^{(-)}\} \cup \{z \# x \mid z \in \Sigma^n - \{1^n\}, x \in \Sigma_\#^*\} \cup \{z \mid z \in \Sigma^n\}$, where $\Sigma_\# = \Sigma \cup \{\#\}$. We set $K = \bigcup_{n \in \mathbb{N}} K_n^{(+)}$ and $K^c = \bigcup_{n \in \mathbb{N}} K_n^{(-)}$. It then follows that $K \cup K^c = \Sigma_\#^*$ and $K \cap K^c = \varnothing$; thus, K^c coincides with the complement $\overline{K}$ of K. We define $m'(w) = n$ if $w = 1^n \# x$ for a certain

$x \in L_n^{(+)} \cup L_n^{(-)}$ and $m'(w) = |w|$ otherwise. Note that m' is log-space computable and also polynomially honest. Since $(L_n^{(+)}, L_n^{(-)})$ is solved by M_n for each $n \in \mathbb{N}$, K is solvable nondeterministically in time $(|x|m(x))^{O(1)}$ and space $O(\log m(x))$ with the use of an appropriate polynomial-size advice function. Thus, (K, m') belongs to para-NL/poly $\cap$ PHSP.

The assumption para-NL/poly $\cap$ PHSP $\subseteq$ para-UL/poly makes (K, m') fall into para-UL/poly. Take an advice function h and a UTM N that solve K in time $(|x|m'(x))^{O(1)}$ and space $O(\log m'(x))$. Consider the algorithm that, on input x with index $n \in \mathbb{N}$, generate both $1^n \# x$ and $h(|x|)$ and then run N on $(1^n \# x, h(|x|))$ to produce an outcome. An appropriate unambiguous 2nfa, say, N_n' can realize this algorithm since we can store $h(|x|)$ as a series of inner states. We thus conclude that N_n' solves $(L_n^{(+)}, L_n^{(-)})$. Therefore, $\mathcal{L}$ belongs to 2U.   □

This completes the proof of Theorem 9.   □

4.2   Case of No Ceiling Restriction

We turn our attention to the case of no ceiling restriction. Kapoutsis [8] demonstrated that $2N \subseteq 2D$ iff $2N/\text{supexp} \subseteq 2D$, where supexp denotes the set of all super-exponentials on $\mathbb{N}$. Similarly, we claim that, in our unambiguity/fewness setting, the case of no ceiling restriction on families of promise problems is equivalent to the case of supexp-ceiling.

Proposition 13. *For any pair $\mathcal{C}$ and $\mathcal{D}$ of classes taken from $\{D, \text{ReachU}, \text{ReachFewU}, \text{ReachFew}, U, \text{FewU}, \text{Few}, N\}$, it follows that $2\mathcal{C} \subseteq 2\mathcal{D}$ iff $2\mathcal{C}/\text{supexp} \subseteq 2\mathcal{D}$.*

For a finite automaton M, $L(M)$ means the set of all strings accepted by M. In the following proof, for simplicity, we assume that 2nfa's make their tape head return to the start cell when they halt.

Proof Sketch of Proposition 13. Since $2\mathcal{C}/\text{supexp} \subseteq 2\mathcal{C}$, it is obvious that $2\mathcal{C} \subseteq 2\mathcal{D}$ implies $2\mathcal{C}/\text{supexp} \subseteq 2\mathcal{D}$. In what follows, we intend to prove the converse. Assume that $2\mathcal{C}/\text{supexp} \subseteq 2\mathcal{D}$. Our goal is to show that $2\mathcal{C} \subseteq 2\mathcal{D}$. Let $f(n)$ denote any function in supexp. It follows that, for any polynomial q, $f(n) > 2^{q(n)}$ holds for all but finitely many $n \in \mathbb{N}$. Let $\mathcal{L} = \{(L_n^{(+)}, L_n^{(-)})\}_{n \in \mathbb{N}}$ be any family of promise problems in $2\mathcal{C}$ and take a family $\mathcal{N} = \{N_n\}_{n \in \mathbb{N}}$ of polynomial-size 2nfa's that solves $\mathcal{L}$, where every N_n must satisfy the condition imposed on underlying machines of $2\mathcal{C}$. Let $N_n = (Q_n', \Sigma, \{\triangleright, \triangleleft\}, \delta_n', q_{0,n}', Q_{acc,n}', Q_{rej,n}')$ and let q denote a polynomial satisfying $|Q_n'| \leq q(n)$ for all $n \in \mathbb{N}$.

We expand $(L_n^{(+)}, L_n^{(-)})$ to the "language" $L(N_n)$ induced by N_n. Note that $L_n^{(+)} \subseteq L(N_n)$ and $L_n^{(-)} \subseteq \overline{L(N_n)}$. For convenience, we write $N_n(x)$ to denote the outcome (i.e., acceptance or rejection) of N_n on input x. From this language $L(N_n)$, we define another promise problem $(K_n^{(+)}, K_n^{(-)})$ by setting $K_n^{(+)} = \{x \in L(N_n) \mid |x| \leq f(n)\}$ and $K_n^{(-)} = \{x \in \overline{L(N_n)} \mid |x| \leq f(n)\}$. Consider the family

$\mathcal{K} = \{(K_n^{(+)}, K_n^{(-)})\}_{n \in \mathbb{N}}$. Since $\mathcal{K}$ has an $f(n)$-ceiling, $\mathcal{K}$ must be in $2\mathcal{C}/f(n)$, which is further included in $2\mathcal{C}/\text{supexp}$ since $f \in \text{supexp}$.

By our assumption, $\mathcal{K}$ belongs to $2\mathcal{D}$. Take a family $\mathcal{M} = \{M_n\}_{n \in \mathbb{N}}$ of 2nfa's with $M_n = (Q_n, \Sigma, \{\triangleright, \triangleleft\}, \delta_n, q_{0,n}, Q_{acc,n}, Q_{rej,n})$ that solves $\mathcal{K}$, where $|Q_n| \leq p(n)$ holds for a fixed polynomial p independent of n and each M_n satisfies the condition imposed for $2\mathcal{D}$. Notice that M_n can take strings of arbitrary lengths as its inputs. However, for any x of length at most $f(n)$, $M_n(x)$ coincides with $N_n(x)$. We then compare between the behaviors of M_n and N_n. Define $r(n) = 2(p(n)^2 + q(n)^2)$ for all $n \in \mathbb{N}$, which is a polynomial satisfying $2(|Q_n|^2 + 2|Q'_n|^2) \leq r(n)$. We then claim the following statement concerning the set $A = \{n \in \mathbb{N} \mid L(M_n) \neq L(N_n)\}$.

Claim 14 $A \subseteq \{n \in \mathbb{N} \mid 2^{r(n)} \geq f(n)\}$.

Finally, we modify M_n into M'_n in order to solve $(L_n^{(+)}, L_n^{(-)})$. By the choice of f, $\{n \in \mathbb{N} \mid 2^{r(n)} \geq f(n)\}$ is a finite set, and thus A is also a finite set. Let c denote the largest number in A. Given any $n \in \mathbb{N}$, if $n \in A$, then we take a 1dfa that exactly simulates N_n with $2^{O(|Q'_n|)}$ inner states and we set M'_n to be this 1dfa; otherwise, we define M'_n to be exactly M_n. Since $n \leq c$, the state complexity of M'_n is upper-bounded by a certain constant, independent of n. Therefore, M'_n correctly solves $(L_n^{(+)}, L_n^{(-)})$, as requested.   □

5   A Discussion and Open Problems

In this work, we have studied the computational complexities of various families of promise problems solved by nonuniform families of polynomial-size nondeterministic finite automata with unambiguity/fewness conditions on their accepting computation paths. In particular, following [12], we have introduced six classes of such families. When tape heads of underlying finite automata are limited to move in only one direction, we have proved that most of those classes are distinct from each other. On the contrary, when the tape heads are allowed to move in both directions, four of the six classes have been shown to collapse. All those results have been illustratively summarized in Fig. 1. As for problems left untold in this work, we wish to list two relevant topics for future work.

It is not known that 1ReachFew ⊆ 1FewU. We have shown that 2N/poly coincides with 2U/poly, but this does not seem to imply the collapse of 2U/poly down to 2ReachFew/poly (or 2ReachFewU/poly or even 2ReachU/poly). Does such a collapse actually occur?

When we change polynomial ceilings to exponential ceilings, for instance, is it true that 2N/exp ⊆ 2U? Is it also true that 2ReachFew/exp ⊆ 2ReachFewU?

Recently, as a natural extension of finite automata, nonuniform families of pushdown automata were studied in [20]. It is unknown that 2N ⊆ 2DPD. For a more weak family, such as 2ReachFewU, is it true that 2ReachFewU ⊆ 2DPD?

References

1. Allender, E.W.: The complexity of sparse sets in P. In: Selman, A.L. (ed.) Structure in Complexity Theory. LNCS, vol. 223, pp. 1–11. Springer, Heidelberg (1986). https://doi.org/10.1007/3-540-16486-3_85
2. Allender, E., Rubinstein, R.: P-printable sets. SIAM J. Comput. **17**, 1193–1202 (1988)
3. Berman P., Lingas, A.: On complexity of regular languages in terms of finite automata, Report 304, Institute of Computer Science, Polish Academy of Science, Warsaw (1977)
4. Bourke, C., Tewari, R., Vinodchandran, N.V.: Directed planar reachability is in unambiguous log-space. ACM Trans. Comput. Theory **1**, 1–17 (2009)
5. Buntrock, G., Jenner, B., Lange, K., Rossmanith, P.: Unambiguity and fewness for logarithmic space. In: the Proceedings of FCT1991, LNCS vol. 529, pp. 168–179 (1991)
6. Kapoutsis, C.A.: Size complexity of two-way finite automata. In: Diekert, V., Nowotka, D. (eds.) DLT 2009. LNCS, vol. 5583, pp. 47–66. Springer, Heidelberg (2009). https://doi.org/10.1007/978-3-642-02737-6_4
7. Kapoutsis, C.A.: Minicomplexity. J. Automat. Lang. Combin. **17**, 205–224 (2012)
8. Kapoutsis, C.A.: Two-way automata versus logarithmic space. Theory Comput. Syst. **55**, 421–447 (2014)
9. Kapoutsis, C.A., Pighizzini, G.: Two-way automata characterizations of L/poly versus NL. Theory Comput. Syst. **56**, 662–685 (2015)
10. Karp, R.M., Lipton, R.J.: Tring machines that take advice. L'Enseigrement Mathématique **28**, 191–209 (1982)
11. Li, M., Vitányi, P.: An Introduction to Kolmogorov Complexity and Its Applications. Third edition. Springer, Cham (2008). https://doi.org/10.1007/978-0-387-49820-1
12. Pavan, A., Tewari, R., Vinodchandran, N.V.: On the power of unambiguity in logspace. Comput. Complex. **21**, 643–670 (2012)
13. Reinhardt, K., Allender, E.: Making nondeterminism unambiguous. SIAM J. Comput. **29**, 1118–1131 (2000)
14. Sakoda, W.J., Sipser, M.: Nondeterminism and the size of two-way finite automata. In: the Proceedings of STOC 1978, pp. 275–286 (1978)
15. Valiant, L.: The relative complexity of checking and evaluating. Inf. Process. Lett. **5**, 20–23 (1976)
16. Yamakami, T.: The 2CNF Boolean formula satisfiability problem and the linear space hypothesis. In: the Proceedings of MFCS 2017, LIPIcs, vol. 83, 1–14 (2017). arXiv preprint arXiv:1709.10453
17. Yamakami, T.: State complexity characterizations of parameterized degree-bounded graph connectivity, sub-linear space computation, and the linear space hypothesis. Theor. Comput. Sci. **798**, 2–22 (2019)
18. Yamakami, T.: Nonuniform families of polynomial-size quantum finite automata and quantum logarithmic-space computation with polynomial-size advice. Inform. Comput. **286**, 104783 (2022). A preliminary version appeared in the Proceedings of LATA 2019, LNCS, vol. 11417, pp. 134–145 (2019)
19. Yamakami, T.: Relativizations of nonuniform quantum finite automata families. In: the Proceedings of UCNC 2019, LNCS, vol. 11493, pp. 257–271 (2019)
20. Yamakami, T.: Parameterizations of logarithmic-space reductions, stack-state complexity of nonuniform families of pushdown automata, and a road to the LOGCFL⊆LOGDCFL/poly question. arXiv preprint arXiv:2108.12779 (2021)

The Stochastic Arrival Problem

Thomas Webster[✉]

University of Edinburgh, Edinburgh, UK
thomas.webster@ed.ac.uk

Abstract. We study a new modification of the Arrival problem, which allows for nodes that exhibit random as well as controlled behaviour, in addition to switching nodes. We study the computational complexity of these extensions, building on existing work on Reachability Switching Games. In particular, we show for versions of the arrival problem involving just switching and random nodes it is PP-hard to decide if their value is greater than a half and we give a PSPACE decision algorithm.

Keywords: Arrival · Markov chains · Reachability Switching Games · MDPs · Simple stochastic games

1 Introduction

Arrival is a simple to describe decision problem defined by Dohrau, Gärtner, Kohler, Matoušek and Welzl [3]. In simplistic terms, it asks whether a train moving along the vertices of a given directed graph, with n vertices, will eventually reach a given target vertex, starting at a given start vertex. At each vertex, v, the train moves deterministically, based on a given listing of outgoing edges of v, taking the first out-edge, then the second, and so on, as it revisits that vertex repeatedly, until the listing is exhausted after which it restarts cyclically at the beginning of the listing of outgoing edges again. This process is known as "switching" and can be viewed as a deterministic simulation of a random walk on the directed graph. It can also be viewed as a natural model of a state transition system where a local deterministic cyclic scheduler is provided for repeated transitions out of each state.

Dohrau et al. showed this Arrival decision problem lies in the complexity class NP ∩ coNP, but it is not known to be in P. There has been a lot of recent work, showing that a search version of the Arrival problem lies in sub-classes of TFNP including PLS [10], CLS [6], and UniqueEOPL [5], as well as showing that Arrival is in UP ∩ coUP [6]. There has also been work on lower bounds, including PL-hardness and CC-hardness [11]. Further recent work by Gärtner et al. [7] gives an algorithm for Arrival with running time $2^{\mathcal{O}(\sqrt{n}\log(n))}$, the first known sub-exponential algorithm. In addition, they give a polynomial-time algorithm for "almost acyclic" instances.

The complexity of Arrival is particularly interesting in the context of other games on graphs, such as Condon's simple stochastic games, mean-payoff games,

A. W. Lin et al. (Eds.): RP 2022, LNCS 13608, pp. 93–107, 2022.
https://doi.org/10.1007/978-3-031-19135-0_7

and parity games [1,9,14], for which the two-player variants are known to be in NP ∩ coNP, whereas the one-player variants have polynomial time algorithms. Arrival however is a zero-player game which has no known polynomial time algorithm and furthermore it was shown by Fearnley et al. [4] that a one-player generalisation of arrival is in fact NP-complete, in stark contrast to these two-player graph games.

Further generalisations of Arrival to Reachability Switching Games were considered, adding player controlled nodes to the game, by Fearnley, Gairing, Mnich and Savani [4]. We provide a further generalisation, by introducing probabilistic nodes, out of which we have random transitions according to a given probability distribution, thus combining the elements of Fearnley et al. [4] and those of Condon's [1], by allowing a mixture of randomisation, switching, and controlled or game behaviour.

Some of our main results consider a mixture of switching and randomisation. In this case we show there is an exponential upper bound on the expected termination time of such a switching run. We also show that deciding whether the value is greater than 0 (or equal to 1 resp.) is complete for NP (resp. coNP) and that the quantitative decision problem is both hard for PP, under many-one (Karp) reductions, and contained in PSPACE thus showing it is harder than the single player switching games of Fearnley et al. [4]. We also give hardness results for the natural generalisation with players, showing these are hard for PSPACE. Some simpler upper bounds follow from viewing these as succinctly presented instances of MDPs, or Condon's simple stochastic games. A full summary of our complexity results (and prior complexity results) can be found in Table 1.

Due to space limits, most proofs are relegated to the full version of the paper.

2   Preliminaries

Our arrival instances represent a reachability problem in a given directed graph, $G = (V, E)$, with given start and target vertices $s, t \in V$, and where the nodes V are partitioned into different types according to a given partition $\mathcal{V}$, with nodes of each type having slightly different behaviour. We use $d_{\text{in}}(v)$ and $d_{\text{out}}(v)$ to represent the in-degree and out-degree of a vertex v in a directed graph. Four distinct types of nodes may be contained in $\mathcal{V}$:

- **Probabilistic nodes** - We denote the set of probabilistic nodes by $V_R \in \mathcal{V}$, and we require a probability distribution, P, to be given on their outgoing edges. These are sometimes also called as random, stochastic or nature nodes.
- **Switching nodes** - We call the set of switching nodes $V_S \in \mathcal{V}$, and require an ordering, Ord, to be given on their outgoing edges.
- **Max Player nodes** - We call the set of max player nodes $V_1 \in \mathcal{V}$ at which choices are controlled by a player aiming to reach t. These are also referred to as player 1 nodes.
- **Min Player nodes** - We call the set of min player nodes $V_2 \in \mathcal{V}$ at which choices are controlled by a player aiming to avoid t. These are also referred to as player 2 nodes.

Our instances then have the following structure.

Definition 1. *An instance of an arrival problem has the following signature* $(V, E, s, t, \mathcal{V}, P, Ord)$ *where:*

- *(V, E) is a finite directed graph.*
- *$s, t \in V$. s is called the* start *and t the* target *node.*
- *For all $v \in V$, we require $d_{out}(v) \geq 1$, and we allow self loop edges of the form (v, v).*
- *For t we require $(t, v) \in E \implies v = t$, i.e. the only out-edge at the target is a self-loop.*
- *$\mathcal{V} \subseteq \mathcal{P}(V)$ is a partition of the vertices of $V - \{t\}$ into different node types. Often we will take $\mathcal{V} = \{V_R, V_S, V_1, V_2\}$, omitting empty sets, with each of these sets as described above.*
- *A function $P : V_R \times V \to [0, 1]$ with the properties that for any $v \in V_R$ we have $\sum_{w \in V} P(v, w) = 1$ and where $P(v, w) > 0$ if and only if $(v, w) \in E$.*
- *A function $Ord : V_S \to V^+$ from switching nodes to a finite sequence of vertices. We require that, for $v \in V_S$, $(v, w) \in E$ if and only if there exists an i such that $w = Ord(v)_i$. So, every outgoing edge from v is "used" in $Ord(v)$, but can be used more than once.*

Given such a model, we wish to define a play of the game. To do so we first need to define the current state. Due to how switching nodes work we will also include the current positions of those nodes into our game state.

Definition 2. *Given a set of switching nodes V_S the* current switching node position *is a function $q : V_S \to \mathbb{N}_0$, i.e., a function from vertices to natural numbers, where we require that $\forall v \in V_S$, $q(v) < |Ord(v)|$. We call the set of all such position functions Q. If there are no switching vertices then Q is a singleton containing only the empty function.*

Definition 3. *A* state *of the game consists of an ordered pair $(v, q) \in V \times Q$ with $v \in V$ denoting the current vertex, and $q \in Q$, denoting the current position of the switching nodes (Definition 2). Thus we call the set $V \times Q$ our* state space.

Now that we have a state space we can define valid transitions between states.

Definition 4. *We let* $Valid : V \times Q \to \mathcal{P}(V \times Q)$ *be the function defined as follows:*

- *For $v \in V_S$ and any $q \in Q$, where by definition $q : V_S \to \mathbb{N}_0$, we define $Valid(v, q) := \{(u, q')\}$, where u and q' are defined as follows:*
  - *Suppose $Ord(v) = (u_0, \ldots, u_{k-1})$. We let $u := u_{q(v)}$. Note that this is well defined, i.e., $0 \leq q(v) < |Ord(v)| = k$, because (v, q) is a state.*
  - *For $x \in V_S$ with $x \neq v$ we let $q'(x) := q(x)$.*
  - *Furthermore, we let $q'(v) := (q(v) + 1 \mod k)$.*
- *For $v \in V_1 \cup V_2$ and any $q \in Q$, we let $Valid(v, q) := \{(u, q) : (v, u) \in E\}$.*
- *For $v \in V_R$ and any $q \in Q$ we let $Valid(v, q) := \{(u, q) : P(v, u) > 0\}$*

We call a transition from a state (v, q) to a state $(u, q') \in Valid(v, q)$ valid, and otherwise we call it invalid.

It follows directly from the definitions that for any state (v, q), $Valid(v, q) \neq \emptyset$.

We call an infinite sequence $\pi = (v_0, q_0)(v_1, q_1)(v_2, q_2) \cdots \in (V \times Q)^\omega$ over the state space $V \times Q$ a *play* if for every $i \in \mathbb{N}_0$ we have $(v_{i+1}, q_{i+1}) \in Valid(v_i, q_i)$. We use Ω to denote the set of all (infinite) plays. A *partial play* of the game is a finite initial prefix $w \in (V \times Q)^*$ of a play. For a partial play w, we define its *basic cylinder*, $\mathsf{C}(w) \subseteq w(V \times Q)^\omega$, as the set of all plays with w as an initial segment. We use $\Pi \subseteq (V \times Q)^*$ to denote the set of all finite partial plays. We say a play π is *winning* for player 1 if there exists some index i with $\pi_i = (t, q)$. Otherwise, it is a losing play (winning for player 2).

It follows from known results, namely, deterministic memoryless determinacy of simple stochastic games ([1]), that for all our generalised arrival games it suffices to consider deterministic "essentially memoryless" strategies for a player i given by $Strat_i : (V_i \times Q) \to V$, which ignore the history in a partial play π, and only considers the current state (v, q) in order to choose (deterministically) a move to the next vertex, v', such that $(v', q) \in Valid(v, q)$. (Note that switching positions only change during transitions out of switching nodes.) Indeed, we can view our instances of generalised arrival as defining exponentially larger simple stochastic games over the state space $V \times Q$, because of the deterministic way the switching position q updates with each transition.

Fixing a start state s, and strategies σ_1 and τ_2 for the two players, naturally determines a probability space $(\Omega, \mathcal{F}, \mathbb{P}_{\sigma_1, \tau_2})$ on the set Ω of (infinite) plays. Here $\mathcal{F}$ denotes the Borel σ-algebra of events generated by the set of basic cylinders $\{\mathsf{C}(w) \mid w \in \Pi\}$, and $\mathbb{P}_{\sigma_1, \tau_2}$ denotes the probability measure defined on events in $\mathcal{F}$ uniquely determined by probabilities of basic cylinders, which are defined inductively in the standard way, starting with the base case given by $\mathbb{P}(\mathsf{C}((s, q_0))) := 1$, where by definition $q_0(v) := 0$ for all $v \in V_S$. In other words, all plays begin, with probability 1, with state (s, q_0) as the initial state.

Definition 5. *Given an instance* $G = (V, E, s, t, \{V_R, V_S, V_1, V_2\}, P, Ord)$ *we define the* value *of the instance as follows. Let* $Reach \in \mathcal{F}$ *be the event* $Reach := \{\pi = (s, q_0)(v_1, q_1)(v_2, q_2) \ldots \in \Omega : \exists i \in \mathbb{N}_0, v_i = t\}$ *and let* σ_1 *and* τ_2 *range over strategies for each player:*

$$val(G) := \max_{\sigma_1} \min_{\tau_2} \mathbb{P}_{\sigma_1, \tau_2}(Reach)$$

We may sometimes refer to the value $val(G)$ *as the "winning probability" (for player 1).*

It follows from known results for simple stochastic games that these games are determined, meaning that $val(G) = \min_{\tau_2} \max_{\sigma_1} \mathbb{P}_{\sigma_1, \tau_2}(Reach)$ and that these maxima and minima are obtained.

We generalise of the notion of a "hopeful edges" of [3, Definition 3]:

Definition 6. *Given an instance $G := (V, E, s, t, \mathcal{V}, P, Ord)$ we say a vertex $v \in G$ is hopeful if Player 1 can win the reachability game $(V, E, v, t, \{V_1', V_2\})$, where $V_1' := V_R \cup V_S \cup V_1$ and v is our start vertex. We call an edge $(v, w) \in E$ a hopeful edge if w is a hopeful vertex. A vertex or edge which isn't hopeful is called* dead.

We note that we can decide whether $v \in G$ is hopeful in NL if we have no player 2 nodes in G, and otherwise in P by solving the 2-player reachability game. We now define different versions of the computational problems we wish to study, using a common notation. We use a subset $B \subseteq \{R, S, 1, 2\}$ to denote the different kinds of nodes that are present in the instances for the problem in question.

Definition 7. *For a subset $B \subseteq \{R, S, 1, 2\}$, given an instance structure $G = (V, E, s, t, \{V_\sigma : \sigma \in B\}, P, Ord)$, we define the following associated decision problems. Let $val(G)$ be the value of the underlying arrival/reachability game associated with G, and let $p \in (0, 1)$ be a (rational) probability given as part of the input. We define three variants of quantitative and qualitative B-Arrival decision problems that we wish to study:*

- *B-Arrival-Quant: Decide whether $val(G) > p$.*
- *B-Arrival-Qual-0: Decide whether $val(G) > 0$.*
- *B-Arrival-Qual-1: Decide whether $val(G) = 1$.*

The original arrival problem studied in [3] corresponds to the above definition with $B = \{S\}$. Reachability Switching Games defined in [4] correspond to $B = \{S, 1\}$ and $B = \{S, 1, 2\}$. Taking $B \subseteq \{R, 1, 2\}$ corresponds to Markov Chains, Markov Decision Processes and Simple Stochastic Games.

We note that when $R \notin B$ these problems all coincide, since in that case $val(G) \in \{0, 1\}$ and such instances constitute an (exponentially large) deterministic reachability game. In such a case we use B-Arrival to refer to the problem of deciding if $val(G) = 1$. Several of these problems have previously known complexity. Throughout this work we aim to show complexity results for the cases when $R \in B$. When referring to an instance of some variant of the above arrival problems, with node types B, we use the expression "instance of a generalised B-arrival problem". It is not hard to show:

Proposition 1. *Given an instance of a generalised B-arrival problem $G = (V, E, s, t, \{V_\sigma : \sigma \in B\}, P, Ord)$, with $R \in B$, and given any rational $p \in (0; 1)$, the decision problem B-Arrival-Quant is polynomial-time equivalent to B-Arrival-Quant where $p = 1/2$.*

Hence we will use B-Arrival-Quant to refer to the quantitative arrival problem when $p = \frac{1}{2}$, and it suffices to only consider this quantitative decision problem. The complexity status of the various different arrival problems, including the results established in this paper, is summarized in Table 1, with references to the original works, or to specific results in this paper that establish it.

While Fearnley et al. do not explicitly consider the $\{S, 2\}$-Arrival problem in [4] we are able to deduce NP-completeness using their results and our generalised notion of hopefulness.

Table 1. Complexity of `Arrival` variants with different node types.

Problem name	Known complexity	Cite
{S}-Arrival	PL-hard, CC-hard (explicit input)	[11]
	P-hard (succinct input)	[4]
	in UEOPL, CLS,PLS, UP ∩ coUP	[3,6]
{S, 1}-Arrival	NP-complete	[4]
{S, 2}-Arrival	NP-complete	Proposition 2
{S, 1, 2}-Arrival	PSPACE-hard in EXPTIME	[4] [4]
{R, S}-Arrival-Qual-0	NP-complete	Theorem 2
{R, S}-Arrival-Qual-1	coNP-complete	Theorem 4
{R, S}-Arrival-Quant	PP-hard, in PSPACE	Theorem 6, Theorem 5
{R, S, 1}-Arrival-Qual-0	NP-Theorem complete	2
{R, S, 1}-Arrival-Qual-1	coNP-hard, in EXPTIME	Theorem 4
{R, S, 1}-Arrival-Quant	PSPACE-hard, in EXPTIME	Theorem 1
{R, S, 2}-Arrival-Qual-0	equiv {S, 1, 2}-Arrival	Theorem 3
{R, S, 2}-Arrival-Qual-1	in EXPTIME	Proposition 5
{R, S, 2}-Arrival-Quant	PSPACE-hard, in EXPTIME	Corollary 2
{R, S, 1, 2}-Arrival-Qual-0	equiv {S, 1, 2}-Arrival	Theorem 3
{R, S, 1, 2}-Arrival-Qual-1	in NEXPTIME ∩ coNEXPTIME	Proposition 5
{R, S, 1, 2}-Arrival-Quant	PSPACE-hard, in NEXPTIME ∩ coNEXPTIME	Theorem 1, Proposition 5

Proposition 2. *The* {S, 2}-Arrival *problem is NP-complete.*

We need a convenient notation for drawing instances of arrival diagrammatically. To do so we use the shapes shown in Fig. 1a to distinguish the different node types. We also make use of gadgets, shown in dashed lines, which are repeated pieces of smaller graphs performing a specific function. Gadgets are shown with entry and exit ports and are permitted to contain other gadgets in a non-recursive way. At probabilistic nodes we assume there is a uniform distribution over outgoing edges, otherwise we label each edge with the probability assigned to it. At switching nodes we label each outgoing edge with numbers such that if $Ord(x) = u_0 \ldots u_k$ we label an edge (x, y) with all the indices, i, such that $u_i = y$. For instance in Fig. 1b we have $k = a+1$ and $Ord(x) = y \ldots yz$, with a consecutive y's.

2.1   Preliminary Results

We may assume Arrival instances have simplified forms and that any instance may be transformed in polynomial-time to an equivalent simplified form. In our simplified form we have two distinguished vertices t and d, with a single self-loop edge. Every other $v \in V \setminus \{t, d\}$ has $d_{\text{out}}(v) = 2$ and $(v, v) \notin E$. For $v \in V$ and $(v, w) \in E$ we have $P(v, w) = \frac{1}{2}$ and for $v \in V_S$ we have $|Ord(v)| = 2$ and there exists functions $s_0, s_1 : V_S \to V$ with $(v, s_0(v)), (v, s_1(v)) \in E$, $Ord(v) = s_0(v)s_1(v)$ and $s_0(v) \neq s_1(v)$.

We may also view a generalised Arrival instance, G, as concise ways of specifying a expanded (exponentially larger) game, $Exp(G)$, without switching. These results are shown in the full version and the construction is analogous to ([4,

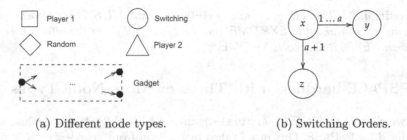

(a) Different node types.   (b) Switching Orders.

Fig. 1. Pictorial representations of B-Arrival instances.

Lemma 4.6]), reducing a 2-player reachability switching game to an exponentially large reachability game. Using this fact we can establish lower bounds on how close the value of such an instance can be to zero, without being equal to zero. Namely, if $val(G)$ is not 0, then, $val(G) = \Omega(2^{2^{-n}})$ where n is our instance bit encoding size.

Corollary 1. *The value of an instance G of a generalised B-arrival problem is a rational number $val(G) := \frac{p}{q}$ which in lowest terms has $0 \leq p, q \leq 4^k$ with $k = 2n(|V| \times M^{|V_S|})$ with $M = \max_{v \in V_S} |Ord(v)|$.*

However we can show that we can actually obtain a value of this small magnitude, even in the case where we only have $B = \{R, S\}$.

Proposition 3. *For any positive integer n, we can construct an instance G of the generalised B-arrival problem containing node types $B = \{R, S\}$, such that G has encoding size $O(n)$, and such that $val(G)$ is a positive value that is at most $\frac{1}{2^{2^n}}$.*

We note that, just as in the case of simple stochastic games, we could force these games to terminate, i.e., reach either the target t or dead end d, by modifying them by applying a small discount, ending the game with a small probability after each step. However, unlike the situation with simple stochastic games, even applying a very small discount of the form $\frac{1}{2^{poly(n)}}$ can change the value of the game drastically (taking a value close to 1 down to a value close to zero). We can however use Proposition 3 to reduce a version of the quantitative B-arrival problem with greater than or equals to the strict inequality decision problem:

Proposition 4. *Given an instance of a generalised B-arrival problem $G = (V, E, s, t, \{V_\sigma : \sigma \in B\}, P, Ord)$, with $R \in B$, and given any rational $p \in (0, 1)$, deciding whether $val(G) \geq p$ is polynomial-time equivalent to B-Arrival-Quant where $p = 1/2$, i.e., to deciding whether $val(G) > 1/2$.*

We can also see from interpreting these models as succinct representations of exponentially large Markov chains, MDPs, and simple stochastic games, respectively, that we have the following simple upper bounds on these problems.

Proposition 5. *The* $\{R,S,1\}$-Arrival-Quant *and* $\{R,S,2\}$-Arrival-Quant *problems are contained in EXPTIME and the* $\{R,S,1,2\}$-Arrival-Quant *is contained in* NEXPTIME $\cap$ coNEXPTIME.

3   PSPACE-hardness with Three or More Node Types

Here we show that $\{R,S,1\}$-Arrival-Quant and $\{R,S,2\}$-Arrival-Quant are both hard for PSPACE. Our proof takes inspiration from Fearnley et al.'s proof of PSPACE-hardness for $\{S,1,2\}$-Arrival ([4, Theorem 4.3]), but requires some new tricks. From these results it trivially follows that $\{R,S,1,2\}$-Arrival-Quant is also PSPACE-hard.

To show the $\{R,S,1\}$-Arrival-Quant is PSPACE-hard we reduce from the SSAT problem as defined by Papadimitriou ([12]):

Definition 8 (SSAT). *Given a 3CNF Boolean formula* $\varphi = C_1 \wedge C_2 \wedge \ldots \wedge C_m$ *with three literals per clause, involving variables* $x_1, \ldots, x_n$, *where* n *is even, we are asked whether there is a choice of Boolean value for* x_1 *such that, for a random choice (with probability of true and false each equal to* $\frac{1}{2}$*) of truth value for* x_2, *there is a choice for* x_3, *etc., so that the probability that* φ *comes out true under these choices is greater than 1/2. We denote this as follows (read* $\Re$ *as "for uniformly random"):*

$$\exists x_1 \Re x_2 \exists x_3 \ldots \Re x_n [\mathbb{P}(\varphi(x_1, \ldots, x_n) = \top) > \frac{1}{2}] \tag{1}$$

By [12, Theorem 2] this problem is PSPACE-complete. Our aim is to take an instance of SSAT and construct an instance $G(\varphi)$ of generalised $\{R,S,1\}$-Arrival with the following property:

$$val(G(\varphi)) = \max_{x_1}[\mathbb{E}_{x_2}[\max_{x_3}[\ldots \mathbb{E}_{x_n}[\chi[\varphi(x_1, \ldots, x_n) = \top]\ldots]] \tag{2}$$

where χ represents the indicator function for an event. With this we can see that $val(G(\varphi)) > 1/2$ if and only if (1) holds. We now outline this construction and show it can be performed efficiently, and that the value is as required.

Given an instance of SSAT with 3CNF φ, n variables and m clauses, we construct the instance $G(\varphi)$ of generalised $\{R,S,1\}$-arrival shown in Fig. 2 where each of the boxes represents the gadgets shown in Figs. 3 and 4, respectively and the values a_i, b_i and D are computable from the formula φ.

We now explain this construction in more detail. Given $\varphi = C_1 \wedge C_2 \wedge \ldots \wedge C_m$, to begin with, in polynomial time we enumerate our n variables as $x_1, \ldots, x_n$ and for each we compute constants $a_i = |\{l \in \{1, \ldots, m\} \mid x_i \in C_l\}|$ and $b_i = |\{l \in \{1, \ldots, m\} \mid \neg x_i \in C_l\}|$. Here a_i is the number of clauses in which the literal x_i appears, and b_i is the number of clauses in which the literal $\neg x_i$ appears. We let $D = \max \bigcup_i \{a_i, b_i\}$ be the maximum number of occurrences of any literal. We divide the game into three phases which correspond to the different nodes in $Ord(start)$: the "assignment" phase, consisting of the time

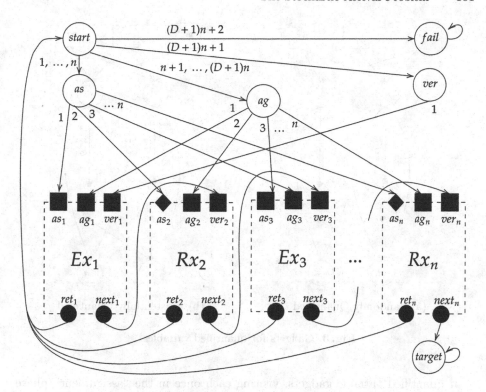

Fig. 2. Control gadget

strictly before the $n + 1$'th visit to the vertex *start* where the switching node takes us to the node *as*, the "agreement" phase, consisting of the time strictly before the $Dn + 1$'th visit to the vertex *start* where the switching node takes us to *ag*, and the "verification" phase consisting of the time afterwards where the switching takes us to either *ver* or *fail*. Each phases has the following objectives:

- **Assignment Phase** - In this phase the player and nature alternate in choosing values of $x_1, \ldots, x_n$ in sequence.
- **Agreement Phase** - In this phase the player must continue to agree with the choices in the "assignment" phase. Each time we visit we go through a list of clauses which our choice of assignment to that variable doesn't satisfy.
- **Verification Phase** - In this phase we verify that the player acted honestly and did agree with the choices in the "assignment" phase by moving through each variable gadget.

These phases correspond to the three distinct entries to each of our quantified variable gadgets and we only use the entrance matching the phase we are in. We use "pass" to refer to a path from an entry to the exit, the "initial pass" is the one made in the "assignment" phase. Our gadgets function like:

- **The Control Gadget.** In this structure shown in Fig. 2 we enforce the phases using the switching behaviour at *start*. The nodes *as* and *ag* cycle through the

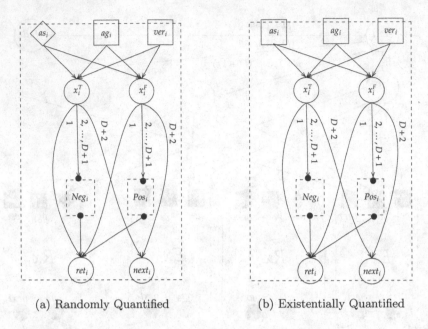

(a) Randomly Quantified   (b) Existentially Quantified

Fig. 3. Gadgets for quantified variables.

n quantified variable gadgets, visiting each once in the "assignment" phase and D times in the "agreement" phase. The node ver finally starts the verification process by moving to ver_1. We note any more visits to $start$ send us to $fail$. We note our quantified gadgets are connected with edges between as and all as_i and between ag and all ag_i, return edges from ret_i to $start$ and a chain of edges going from ver to ver_1, $next_1$ to ver_2,..., and finally $next_n$ to $target$.

- **Quantified Variable Gadget.** We have two variations of this gadget shown in Figs. 3a and b which depend on whether x_i is existentially or randomly quantified in φ, differing only in the node type of as_i. On the initial pass the assignment is chosen by the player or uniformly at random respectively. The three entries correspond to the different phases of the game and we have two exits, ret_i returns back to the $start$ and $next_i$ moves us on to the next variable's verification entry ver_{i+1}, or to $target$ if $i = n$. The nodes x_i^T and x_i^F represent choosing an assignment of the variable x_i on this pass, and the "initial assignment" is the one from the initial pass. The switching behaviour of x_i^T and x_i^F prevents $next_i$ being reached without $D + 2$ visits to one of the two nodes, which forces D visits to the respective Consequence gadget Neg_i or Pos_i.

- **Consequences Gadget.** We have two consequences gadgets for each variable, Neg_i and Pos_i, shown in Figs. 4a and b. Neg_i (resp. Pos_i) enumerates the gadgets for clauses, $C_{j_1}, \ldots, C_{j_{a_i}}$ (resp. $C_{k_1}, \ldots, C_{k_{b_i}}$), where the literal $\neg x_i$ (resp. x_i) appears. When we choose an assignment of true (resp. false)

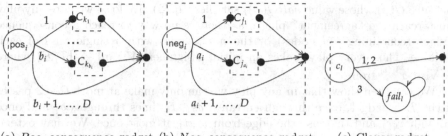

(a) Pos_i consequence gadget   (b) Neg_i consequence gadget   (c) Clause gadget

Fig. 4. Gadgets for Consequences of variable x_i and Clauses C_l

these clauses aren't immediately satisfied by our assignment. As any literal appears in at most D clauses by visiting this gadget D times we are guaranteed to go through each of the contained clause gadgets. If we have $a_i < D$ (resp. $b_i < D$) then any further edges proceed straight to the exit to ensure if we make exactly D passes we visit each clause gadget exactly once.

- **The Clause Gadget.** This is shown in Fig. 4c. Here we check if it is possible to still satisfy a clause. Note we pass through the clause gadget for C_l only in the following situations:
  - From a Neg_i gadget where we have assigned x_i true on this pass and $\neg x_i$ appears in C_l,
  - From a Pos_i gadget where we have assigned x_i false on this pass and x_i appears in C_l,

Thus as a consequence of our truth assignment to x_i it doesn't witness the truth of C_l. Our clause C_l has width 3 and if our assignment is satisfying then we must have at least one of the 3 literals as a witness to the truth of C_l. Thus our gadget acts as a simple counter of the number of literals in the clause which evaluate to false, after 3 passes our switch sends the play to the fail state, because the assignment we have chosen does not satisfy C_l. On the first and second passes the counter is just incremented and we use this gadget to ensure the clause is satisfied.

We can prove that instance $G(\varphi)$ has value $val(G(\varphi))$ satisfying Eq. (2).

We note that this construction remains polynomial in the size of the formula, with the control gadget (Fig. 2) only containing instances of the randomly and existentially quantified variable gadgets, the quantified variable gadgets (Fig. 3) only containing the Consequence gadgets Pos_i and Neg_i and the Consequence gadgets (Figs. 4a and b) only containing Clause Gadgets (Fig. 4c). Further the ret_i exits and all exits of the consequence and clause gadgets may be treated as the node $start$, independent of the index i or l of the gadget, as each has an onward path containing only nodes of out-degree one leading to $start$.

Theorem 1. $\{R, S, 1\}$-Arrival-Quant *is PSPACE-hard.*

Proof (sketch). We prove this by showing the above construction, which can easily be carried out in polynomial time, given a SSAT instance, φ, constructs an instance $G(\varphi)$ whose value $val(G(\varphi))$ satisfies Eq. (2). To do so we note any play must reach the "agreement" phase, as there is no way to reach a consequence gadget (containing $fail$ nodes) or the $next_i$ nodes with a single pass of each variable. Thus every play makes an initial assignment $V : [n] \to \{T, F\}$ where we visit $x_i^{V(i)}$ from as_i.

We can then show that in any play we can only make at most $D + 2$ passes of the Ex_1 gadget, once through entrance as_1, D times through ag_1 and once through ver_1 and thus use the edge from $next_1$ at most once. We may extend this inductively to show in any play we can make at most $D + 2$ passes of any quantified variable gadget and use the $next_i$ exit at most once. We can also show by induction if we reach $target$ we must make exactly $D + 2$ passes of each gadget and use the $next_i$ exit exactly once. To use the $next_i$ exit we must visit one of x_i^T or x_i^F exactly $D + 2$ times.

Firstly we can use this to show in any play that reaches $target$ that the initial valuation V was satisfying. As we make $D + 2$ visits to x_i^T (resp. x_i^F) in the "agreement" phase we must visit exactly one of Neg_i (resp. Pos_i) exactly D times, which means we visit every clause gadget they contain exactly once. If we reach the end of the "agreement" phase then there is at least one edge incoming to each clause gadget that was unused, as there are three incoming edges which can be used at most once each and we can not make three passes of the clause gadget as it has an internal $fail$ state. This lets us show valuation V satisfies φ.

Secondly we can show that under the "agreement strategy", where the player agrees with the initial assignment in the "agreement" and "verification" phases, the play reaches $target$ when V satisfies φ, and by the above we can never reach $target$ otherwise. Thus this strategy is optimal for the player in the "agreement" and "verification" phases.

We then show our value is the maximum over strategies for the "assignment" phase. In this phase we can consider the player and nature playing a game on a binary tree, where the leaves are possible valuations $V : [n] \to \{T, F\}$ and we call a leaf accepting if it's a valuation satisfying φ. At the root the player makes the choice between $V(1) = T$ and $V(1) = F$. On the next level nature randomises between $V(2) = T$ or $V(2) = F$. The player then chooses between $V(3) = T$ or $V(3) = F$, etc. At each stage the player knows the past decisions and maximises their choice with the aim that they reach an accepting leaf, which gives exactly Eq. (2). □

As an immediate corollary we can deduce hardness for $\{R, S, 2\}$-Arrival-Quant.

Corollary 2. $\{R, S, 2\}$-Arrival-Quant *is PSPACE-hard.*

4   The $\{R,S\}$-Arrival Problems

Firstly we give some bounds on the qualitative problems, then we give an interesting bound on the expected number of times we use edges in each play. Finally, for $\{R, S\}$-Arrival-Quant both a PSPACE algorithm and PP-hardness.

We are able to give two easy reductions by creating new instances where we give control of random nodes to player 1 or randomise over player 1 choices, these allow us to deduce NP-completeness for two of our problems.

Theorem 2. $\{R, S\}$-**Arrival-Qual-0**, $\{S, 1\}$-**Arrival** and $\{R, S, 1\}$-**Arrival-Qual-0** *are all poly-time equivalent and NP-complete.*

Theorem 3. $\{R, S, 1, 2\}$-**Arrival-Qual-0**, $\{S, 1, 2\}$-**Arrival** and $\{R, S, 2\}$-**Arrival-Qual-0** *are all poly-time equivalent.*

While the above arguments exploit exchanging player 1 and random nodes, we note that a similar exchange for player 2 is not immediately possible. Consider the case of a cycle of random nodes. Any play must almost surely escape this cycle, however under player 2 control it is optimal to always stay in the cycle.

We now show coNP-hardness of $\{R, S\}$-**Arrival-Qual-1**, by exploiting a construction in [3]. They showed that the $\{S\}$-**Arrival** problem lies in the class NP ∩ coNP by constructing succinct witnesses for the fact that the play does *not* reach the target t, by modifying the graph (such that reachability of t is preserved) introducing a new dead end state d, and showing that exactly one of t or d is reached in any play in the modified graph. Here we show we can use a similar construction to reduce the complement of $\{R, S\}$-**Arrival-Qual-0** to $\{R, S\}$-**Arrival-Qual-1**. We assume (w.l.o.g.), we are working with the instances of Arrival in our simplified form.

Definition 9 (cf. [3, Definition 3]). *Let* $(V, E, s, t, \{V_S, V_R\}, P, Ord)$ *be an instance of generalised* $\{R, S\}$-*arrival. If* $(v, w) \in E$ *is hopeful (Definition 6) we call its* desperation *the length of the shortest directed path from* w *to* t.

We proceed to give our generalised versions of a Lemma in [3], generalised to the randomised setting. We note that it is simple to process our inputs and replace dead edges of the form (v, w) by an edge (v, d) immediately to the dead end.

Definition 10. *Let* G *be an instance of the generalised B-arrival problem and* $e \in E$ *an edge. Define the random variable* T_e *to be the number of traversals of* e *in a run of the instance starting from* s.

Lemma 1. *Let* G *be an instance of the generalised* $\{R, S\}$-*arrival problem in simple form, and let* $e \in E$ *be a hopeful edge of desperation* k *in* G. *Then* $\mathbb{E}[T_e] \leq 2^{k+1} - 1$.

Lemma 1 (which is closely related to [3, Lemma 2]) enables us to bound the expected length of a play by a single exponential in our input $\{R, S\}$-arrival instance size. Note this is despite the fact the $\{R, S\}$-arrival instance succinctly represents an exponentially larger Markov chain, and in general for an exponentially large Markov chain the worst case expected termination (hitting) time can be double-exponential. Note also that by contrast, by Proposition 3, the probability of reaching the target can be double-exponentially small. Using Lemma 1 we construct instances that almost surely terminate and given an instance G construct a new instance G' with $val(G') = 1 - val(G)$. These allow us to show:

Theorem 4. *The* $\{R, S\}$-`Arrival-Qual-1` *problem is* coNP-*complete.*

Theorems 2 and 4 together imply that $\{R, S\}$-`Arrival-Quant` is both NP-hard & coNP-hard. In Theorem 6 we will show a stronger PP-hardness result for $\{R, S\}$-`Arrival-Quant`. As an upper bound, we can show the following:

Theorem 5. *The* $\{R, S\}$-`Arrival-Quant` *problem is in* PSPACE.

Proof (sketch). We can view our instance G as an exponentially larger Markov Chain (MC) with a succinct represented transition probability matrix P. Using suitable preprocessing, we can simplify the model so that the matrix $(I - P)$ is invertible, without altering the probability of reaching the target. We can compute individual bits of the hitting probabilities on such a MC by computing entries of $(I - P)^{-1}$, which can be done in PSPACE, using the fact that an (explicitly given) linear system of equations can be solved in NC2 ([2]). Using these bits we can decide $\{R, S\}$-`Arrival-Quant`. □

We can finally use a construction similar to Theorem 1 to construct a hard instance.

Theorem 6. $\{R, S\}$-`Arrival-Quant` *is PP-hard.*

Proof (sketch). We show this by a reduction from the MAJSAT problem, namely deciding whether, for a given CNF formula $\varphi(x)$ over n variables, the probability, p_φ, that a uniformally random assignment of truth values to the variables x satisfy φ. MAJSAT is PP-complete ([8,13]). We use similar gadgets to those in the proof of Theorem 1, however for our PP-hardness proof for $\{R, S\}$-`Arrival-Quant`, we make a new random assignment on each pass of the variable gadget and use switching nodes to ensure this is the same as past choices. Where we make different assignments to a variable on different passes we move to a state which moves us randomly to *target* or *fail*, thus we only reach the verification phase when we make the same assignment on every pass. Our "verification" phase then checks if all clauses are satisfied. This allows us to distinguish three distinct cases, "invalid random assignment", "valid, unsatisfying assignment" and "valid, satisfying assignment", which we can use to determine if $p_\varphi > \frac{1}{2}$. □

Acknowledgements. Thanks to a prior anonymous reviewer who sketched a proof of Theorem 5, improving on our prior result which only showed that approximation of the $\{R, S\}$-`Arrival` value to within any given desired accuracy $\epsilon > 0$ is in PSPACE.

References

1. Condon, A.: The complexity of stochastic games. Inf. Comput. **96**(2), 203–224 (1992)
2. Csanky, L.: Fast parallel matrix inversion algorithms. SIAM J. Comput. **5**(4), 618–623 (1976). https://doi.org/10.1137/020504

3. Dohrau, J., Gärtner, B., Kohler, M., Matoušek, J., Welzl, E.: ARRIVAL: a zero-player graph game in NP ∩ coNP. In: Loebl, M., Nešetřil, J., Thomas, R. (eds.) A Journey Through Discrete Mathematics, pp. 367–374. Springer, Cham (2017). https://doi.org/10.1007/978-3-319-44479-6_14
4. Fearnley, J., Gairing, M., Mnich, M., Savani, R.: Reachability switching games. Log. Methods Comput. Sci. **17**(2) (2021)
5. Fearnley, J., Gordon, S., Mehta, R., Savani, R.: Unique end of potential line. In: 46th International Colloquium on Automata, Languages, and Programming (ICALP 2019), LIPIcs, vol. 132, pp. 1–15 (2019)
6. Gärtner, B., Hansen, T.D., Hubácek, P., Král, K., Mosaad, H., Slívová, V.: ARRIVAL: next stop in CLS. In: 45th International Colloquium on Automata, Languages, and Programming, LIPIcs, vol. 107, pp. 1–13 (2018)
7. Gärtner, B., Haslebacher, S., Hoang, H.P.: A subexponential algorithm for ARRIVAL. In: 48th International Colloquium on Automata, Languages, and Programming, LIPIcs, vol. 198, pp. 1–14 (2021)
8. Gill, J.T.: Computational complexity of probabilistic turing machines. In: Proceedings of the sixth annual ACM Symposium on Theory of Computing, pp. 91–95. STOC, ACM Press (1974)
9. Jurdzinski, M.: Deciding the winner in parity games is in UP ∩ coUP. Inf. Process. Lett. **68**(3), 119–124 (1998)
10. Karthik, C.S.: Did the train reach its destination: the complexity of finding a witness. Inf. Process. Lett. **121**, 17–21 (2017)
11. Manuell, G.: A simple lower bound for ARRIVAL. arXiv preprint arXiv:2108.06273 (2021)
12. Papadimitriou, C.H.: Games against nature. J. Comput. Syst. Sci. **31**(2), 288–301 (1985)
13. Simon, J.: On some central problems in computational complexity, Ph. D. thesis, Cornell University (1975)
14. Zwick, U., Paterson, M.: The complexity of mean payoff games on graphs. Theor. Comput. Sci. **158**(1–2), 343–359 (1996)

On Higher-Order Reachability Games Vs May Reachability

Kazuyuki Asada[1], Hiroyuki Katsura[2], and Naoki Kobayashi[2(✉)]

[1] Tohoku University, Sendai, Japan
[2] The University of Tokyo, Tokyo, Japan
koba@is.s.u-tokyo.ac.jp

Abstract. We consider the reachability problem for higher-order functional programs and study the relationship between reachability games (i.e., the reachability problem for programs with angelic and demonic nondeterminism) and may-reachability (i.e., the reachability problem for programs with only angelic nondeterminism). We show that reachability games for order-n programs can be reduced to may-reachability problems for order-$(n + 1)$ programs, and vice versa. We formalize the reductions by using higher-order fixpoint logic and prove their correctness. We also discuss applications to higher-order program verification.

1 Introduction

This paper considers the reachability problem for simply-typed, call-by-name higher-order functional programs with integers, recursion, and two kinds of non-deterministic branches (angelic and demonic ones). The problem of solving reachability games (hereafter, simply called the reachability game problem) asks, given a higher-order functional program and a specific control point **succ** of the program, whether there exists a sequence of choices on angelic non-determinism that makes the program reach **succ** no matter what choices are made on demonic non-determinism. Thus, our reachability game problem is just a special case of the notion of two-player reachability games [8], where the game arena is specified as a higher-order functional program. Various program verification problems can be reduced to the reachability game problem. For example, the termination problem, which asks whether a given program terminates for any sequence of non-deterministic choices, is a special case of the reachability game problem, where all the non-deterministic branches are demonic, and all the termination points are expressed by **succ**. The safety verification problem, which asks whether a given program may fall into an error state after some sequence of non-deterministic choices, is also a special case, where all the non-deterministic branches are angelic, and error states are expressed by **succ**.

We establish relations between the reachability game problem and the *may-reachability* problem, a special case of the reachability game problem where all the non-deterministic choices are angelic (hence, may-reachability is a one-player game). We show mutual translations between the reachability game problem for order-n programs and the may-reachability problem for order-$(n+1)$ programs. (Here, the order of a program is defined as the type-theoretic order; the order of a function that takes only integers is 0, and the order of a function that takes an order-0 function is 1, etc.) The translations are size-preserving in the sense that for any order-n program M, one can effectively construct an order-$(n+1)$ program M' such that the answer to the reachability game problem for M is the same as the answer to the may-reachability problem for M', and the size of M' is polynomial in that of M; and vice versa.

The translation from reachability games to may-reachability allows us to use higher-order program verification tools specialized to may-reachability (or, unreachability to error states) such as MoCHi [15] and Liquid types [20] to check a wider class of properties represented as reachability games. Conversely, the translation from may-reachability to reachability games allows us, for example, to use verification tools that can solve reachability games for order-0 programs, such as CHC solvers [5,9,17] to check may-reachability of order-1 programs.

We formalize our translations for μHFL(Z), which is a fragment HFL(Z) [16] without greatest fixpoint operators and modal operators, where HFL(Z) is an extension of Viswanathan and Viswanathan's higher-order fixpoint logic [23] with integers. The use of higher-order fixpoint formulas rather than higher-order programs in the formalization of the translations is justified by the result of Kobayashi et al. [16,24], that there is a direct correspondence between the reachability problem for higher-order programs and the validity problem for the corresponding higher-order fixpoint formulas, where angelic and demonic branches in programs correspond to disjunctions and conjunctions respectively.

The rest of this paper is structured as follows. Section 2 introduces μHFL(Z), and explains its relationship with the reachability problem for higher-order programs. Section 3 gives a reduction from the reachability game problem to may-reachability problem, and Sect. 4 gives a reduction in the opposite direction. Section 5 discusses applications and reports some experimental results. Section 6 discusses related work and Sect. 7 concludes the paper. The proofs and definitions omitted in this paper are found in the longer version [1].

2 μHFL(Z) and Reachability Problems

In this section, we first introduce μHFL(Z), a fragment of higher-order fixpoint logic HFL(Z) [16] (which is in turn an extension of Viswanathan and Viswanathan's higher-order fixpoint logic [23] with integers) without greatest fixpoint operators. We then review the relationship between μHFL(Z) and reachability problems, and state the main theorem of this paper.

2.1 μHFL(Z)

The set of *(simple) types*, ranged over by κ, is given by:

$$\kappa \text{ (types) } ::= \texttt{Int} \mid \tau \qquad \tau \text{ (predicate types) } ::= \star \mid \kappa \to \tau.$$

$$\overline{\mathcal{K}, x:\kappa \vdash_{ST} x:\kappa}$$

$$\dfrac{\mathcal{K} \vdash_{ST} \varphi_1 : \star \quad \mathcal{K} \vdash_{ST} \varphi_2 : \star}{\mathcal{K} \vdash_{ST} \varphi_1 \vee \varphi_2 : \star}$$

$$\dfrac{\mathcal{K} \vdash_{ST} \varphi_1 : \star \quad \mathcal{K} \vdash_{ST} \varphi_2 : \star}{\mathcal{K} \vdash_{ST} \varphi_1 \wedge \varphi_2 : \star}$$

$$\dfrac{\mathcal{K}, x:\tau \vdash_{ST} \varphi:\tau}{\mathcal{K} \vdash_{ST} \mu x^\tau.\varphi : \tau}$$

$$\dfrac{\mathcal{K} \vdash_{ST} \varphi_1 : \tau_2 \to \tau \quad \mathcal{K} \vdash_{ST} \varphi_2 : \tau_2}{\mathcal{K} \vdash_{ST} \varphi_1\varphi_2 : \tau}$$

$$\dfrac{\mathcal{K}, x:\kappa \vdash_{ST} \varphi:\tau}{\mathcal{K} \vdash_{ST} \lambda x^\kappa.\varphi : \kappa \to \tau}$$

$$\dfrac{\mathcal{K} \vdash_{ST} \varphi : \mathrm{Int} \to \tau \quad \mathcal{K} \vdash_{ST} e : \mathrm{Int}}{\mathcal{K} \vdash_{ST} \varphi e : \tau}$$

$$\dfrac{\mathcal{K} \vdash_{ST} e_1 : \mathrm{Int} \quad \mathcal{K} \vdash_{ST} e_2 : \mathrm{Int}}{\mathcal{K} \vdash_{ST} e_1 \leq e_2 : \star}$$

$$\overline{\mathcal{K} \vdash_{ST} n : \mathrm{Int}}$$

$$\dfrac{\mathcal{K} \vdash_{ST} e_1 : \mathrm{Int} \quad \mathcal{K} \vdash_{ST} e_2 : \mathrm{Int}}{\mathcal{K} \vdash_{ST} e_1 + e_2 : \mathrm{Int}}$$

$$\dfrac{\mathcal{K} \vdash_{ST} e_1 : \mathrm{Int} \quad \mathcal{K} \vdash_{ST} e_2 : \mathrm{Int}}{\mathcal{K} \vdash_{ST} e_1 \times e_2 : \mathrm{Int}}$$

Fig. 1. Simple type system for μHFL(Z)

For a type κ, the *order* and *arity* of κ, written $\mathrm{ord}(\kappa)$ and $\mathrm{ar}(\kappa)$ respectively, are defined by: $\mathrm{ord}(\mathrm{Int}) = -1$, $\mathrm{ord}(\star) = 0$, $\mathrm{ord}(\kappa \to \tau) = \max(\mathrm{ord}(\tau), \mathrm{ord}(\kappa) + 1)$, $\mathrm{ar}(\mathrm{Int}) = \mathrm{ar}(\star) = 0$, and $\mathrm{ar}(\kappa \to \tau) = \mathrm{ar}(\tau) + 1$.

The set of μHFL(Z) *formulas*, ranged over by φ, is given by:

$$\varphi \ (\text{formulas}) ::= x \mid \varphi_1 \vee \varphi_2 \mid \varphi_1 \wedge \varphi_2 \mid \mu x^\tau.\varphi \mid \varphi_1\varphi_2 \mid \lambda x^\kappa.\varphi \mid \varphi e \mid e_1 \leq e_2$$
$$e \ (\text{integer expressions}) ::= n \mid x \mid e_1 + e_2 \mid e_1 \times e_2.$$

Intuitively, $\mu x^\tau.\varphi$ denotes the least predicate x of type τ such that $x = \varphi$. We write $\mathtt{true}$ and $\mathtt{false}$ for $0 \leq 0$ and $1 \leq 0$. For a formula φ, the *order* of φ is defined as: $\max(\{0\} \cup \{\mathrm{ord}(\tau) \mid \mu x^\tau.\varphi' \text{ occurs in } \varphi\})$. We call a μHFL(Z) formula φ *disjunctive* if the conjunction $\wedge$ occurs in φ only in the form of $e_1 \leq e_2 \wedge \varphi_1$ (i.e., the left-hand side of φ is a primitive constraint on integers).

We write $\widetilde{\varphi}_{j,\ldots,k}$ for a sequence of formulas $\varphi_j, \ldots, \varphi_k$; it denotes an empty sequence if $k < j$. We often omit the subscript and just write $\widetilde{\varphi}$ for $\widetilde{\varphi}_{j,\ldots,k}$ when the subscript is not important. Similarly, we also write $\widetilde{e}$ and $\widetilde{\kappa}$ for sequences of expressions and types respectively. We use the metavariables α, β, and γ to denote either a formula or an integer expression.

The simple type system for μHFL(Z) formulas is defined in Fig. 1. Henceforth, we consider only well-typed formulas (i.e., formulas φ such that $\mathcal{K} \vdash_{ST} \varphi : \kappa$ for some $\mathcal{K}$ and κ). A formula φ is called a *closed* formula of type κ if $\emptyset \vdash_{ST} \varphi : \kappa$.

For a closed formula φ, we write $[\![\varphi]\!]$ for the semantics of φ. If φ has type $\star$, then $[\![\varphi]\!]$ is either $\top$ (meaning that the formula is valid) or $\bot$ (invalid). Similarly, for a closed expression e, we write $[\![e]\!]$ for the integer value of e. The formal semantics of formulas is found in the full version of this paper [1]. The *validity checking problem* for μHFL(Z) is the problem of deciding whether $[\![\varphi]\!] = \top$, given a closed μHFL(Z) formula φ of type $\star$. Note that the validity checking problem is undecidable.

For closed formulas, the following alternative semantics is sometimes convenient. Let us define the reduction relation $\varphi \longrightarrow \varphi'$ by the following rules.

$$\frac{i \in \{1, 2\}}{E[\varphi_1 \vee \varphi_2] \longrightarrow E[\varphi_i]} \qquad \frac{}{E[\mathbf{false} \wedge \varphi] \longrightarrow E[\mathbf{false}]} \qquad \frac{}{E[(\lambda x.\varphi)\psi] \longrightarrow E[[\psi/x]\varphi]}$$

$$\frac{}{E[\mathbf{true} \wedge \varphi] \longrightarrow E[\varphi]} \qquad \frac{}{E[\mu x.\varphi] \longrightarrow E[[\mu x.\varphi/x]\varphi]} \qquad b = \begin{cases} \mathbf{true} & \text{if } [\![e_1]\!] \leq [\![e_2]\!] \\ \mathbf{false} & \text{otherwise} \end{cases}$$

$$\frac{}{E[(\lambda x.\varphi)e] \longrightarrow E[[e/x]\varphi]} \qquad \frac{}{E[e_1 \leq e_2] \longrightarrow E[b]}$$

Here, E denotes an evaluation context, defined by: $E ::= [\,] \mid E \wedge \varphi \mid E\,\varphi$. We write $\longrightarrow^*$ for the reflexive and transitive closure of $\longrightarrow$. We have the following fact (see, e.g., [22]).

Fact 1. *Suppose* $\vdash_{\mathrm{ST}} \varphi : \star$. *Then,* $[\![\varphi]\!] = \top$ *if and only if* $\varphi \longrightarrow^* \mathbf{true}$.

Due to the fact above, the validity checking problem is equivalent to the problem of deciding whether $\varphi \longrightarrow^* \mathbf{true}$, given a closed μHFL(Z) formula φ of type $\star$.

Example 1. Suppose $\vdash_{\mathrm{ST}} \varphi : \mathrm{Int} \to \star$, and let ψ be the formula $(\mu x^{\mathrm{Int} \to \star}.\lambda y.\varphi\,y \vee \varphi(-y) \vee x(y+1))0$. Then $\psi \longrightarrow^* \mathbf{true}$ just if $\varphi\,n \longrightarrow^* \mathbf{true}$ for some n. Thus, ψ represents $\exists z.\varphi\,z$.

The example above indicates that existential quantifiers on integers are expressible in μHFL(Z). Below, we treat existential quantifiers as if they were primitives.

2.2 Relationship with Reachability Problems

We consider reachability problems for a call-by-name, simply-typed λ-calculus extended with two kinds of non-determinism (■ and □) and a special term **succ**, which represents that the designated target has been reached.[1] The sets of types and terms, ranged over by σ and M respectively, are defined by:

$$\sigma ::= \mathrm{Int} \mid \eta \qquad \eta ::= \mathrm{unit} \mid \sigma \to \eta$$
$$M ::= (\,) \mid \mathbf{succ} \mid x \mid \lambda x.M \mid M_1\,M_2 \mid M\,e$$
$$\mid \mathbf{fix}^\eta(x, M) \mid M_1 \blacksquare M_2 \mid M_1 \square M_2 \mid \mathbf{assume}(e_1 \leq e_2); M.$$

Here, $\mathbf{fix}^\eta(x, M)$ denotes a recursive function x of type η such that $x = M$. The term $M_1 \blacksquare M_2$ denotes a *demonic* choice between M_1 and M_2, where the choice is up to the environment (or, the opponent O of the reachability game), and $M_1 \square M_2$ denotes an *angelic* choice between M_1 and M_2, where the choice is up to the term (or, the player P of the reachability game). The term $\mathbf{assume}(e_1 \leq e_2); M$ first checks whether $e_1 \leq e_2$ holds and if so, proceeds to evaluate M; otherwise aborts the evaluation of the whole term. Using **assume**, we can express a conditional expression **if** $e_1 \leq e_2$ **then** M_1 **else** M_2 as $(\mathbf{assume}(e_1 \leq e_2); M_1) \square (\mathbf{assume}(e_2 + 1 \leq e_1); M_2)$. Henceforth, we consider only terms well-typed in the simple type system (which is standard, hence omitted), where $(\,)$ and **succ** are given type unit.

---

[1] In the context of program verification, we are often interested in (un)reachability to bad states. Thus, in that context, **succ** in this section is actually interpreted as an error state, and the terms "angelic" and "demonic" below are swapped.

The order of a type σ is defined by:

$$\text{ord}(\text{Int}) = -1 \qquad \text{ord}(\text{unit}) = 0 \qquad \text{ord}(\sigma \to \eta) = \max(\text{ord}(\eta), \text{ord}(\sigma) + 1).$$

The order of a term M is defined as the largest order of type η such that M has a subterm of the form $\textbf{fix}^\eta(x, M')$. We write $\text{Int}^n \to \star$ for $\underbrace{\text{Int} \to \cdots \text{Int}}_{n} \to \star$.

For a closed simply-typed term M of type unit, a *play* is a (possibly infinite) sequence of reductions of M. The play is won by the player P if it ends with $\textbf{succ}$; otherwise the play is won by the opponent O. The *reachability game* for M is the problem of deciding which player (P or O) has a winning strategy. For the general notion of reachability games and strategies, we refer the reader to [8]. As a special case of the translation of Watanabe et al. [24] from temporal properties of programs to HFL(Z) formulas, we obtain the following translation $(\cdot)^\dagger$ from reachability games to μHFL(Z) formulas.

$$()^\dagger = \texttt{false} \quad \textbf{succ}^\dagger = \texttt{true} \quad x^\dagger = x \quad (\lambda x.M)^\dagger = \lambda x.M^\dagger \quad (M_1 M_2)^\dagger = M_1^\dagger M_2^\dagger$$

$$(M\,e)^\dagger = M^\dagger\,e \quad (\textbf{fix}(x, M))^\dagger = \mu x.M^\dagger \quad (M_1 \blacksquare M_2)^\dagger = M_1^\dagger \wedge M_2^\dagger$$

$$(M_1 \square M_2)^\dagger = M_1^\dagger \vee M_2^\dagger \quad (\textbf{assume}(e_1 \leq e_2); M)^\dagger = e_1 \leq e_2 \wedge M^\dagger.$$

The following is a special case of the result of Watanabe et al. [24].

Theorem 1 ([24]). *For any closed simply-typed term M of type unit and order k, $M^\dagger$ is a closed μHFL(Z) formula of type $\star$ and order k. The player P wins the reachability game for M, if and only if, $[\![M^\dagger]\!] = \top$.*

Based on the result above, we focus on the validity checking problem for μHFL(Z) formulas, instead of directly discussing the reachability problem. Note that the may-reachability problem (of asking whether, given a closed term M of which all the branches are angelic, there exists a reduction sequence from M to $\textbf{succ}$) corresponds to the validity checking problem for disjunctive μHFL(Z) formulas.

Example 2. Let us consider the following OCaml program.

```
let rec sum x k =
    assert(x>=0); if x=0 then k 0 else sum(x-1)(fun y-> k(x+y))
in sum n (fun r -> assert(r>=n))
```

Suppose we are interested in checking whether the program suffers from an assertion failure. It is modeled as the reachability problem for the term $M_{sum}\,n\,(\lambda r.\textbf{assume}(r < n); \textbf{succ})$, where M_{sum} is:

$$\textbf{fix}(sum, \lambda x.\lambda k.(\textbf{assume}(x < 0); \textbf{succ})$$
$$\square(\textbf{assume}(x = 0); k\,0)\square(\textbf{assume}(x > 0); sum\,(x - 1)\,(\lambda y.k(x + y)))).$$

Here, note that an assertion failure is modeled as $\textbf{succ}$ in our language. By Theorem 1, the above term is reachable to $\textbf{succ}$ just if the (disjunctive) μHFL(Z) formula $\varphi_{ex1} := \varphi_{sum}\,n\,(\lambda r.r < n)$ is valid, where φ_{sum} is:

$$\mu sum.\lambda x.\lambda k.x < 0 \vee (x = 0 \wedge k\,0) \vee (x > 0 \wedge sum\,(x - 1)\,(\lambda y.k(x + y))).$$

The formula φ_{ex1} is valid only if $n < 0$, which implies that the OCaml program suffers from an assertion failure just if $\mathbf{n} < 0$. □

2.3 Main Theorem

The main theorem of this paper is stated as follows.

Theorem 2. *There exist polynomial-time translations $(\cdot)^{\#}$ and $(\cdot)^{\flat}$ between order-n $\mu HFL(Z)$ formulas and order-$(n+1)$ disjunctive $\mu HFL(Z)$ formulas that satisfy: (i) For any order-n closed $\mu HFL(Z)$ formula φ, $\varphi^{\#}$ is an order-$(n+1)$ closed disjunctive $\mu HFL(Z)$ formula such that $[\![\varphi]\!] = [\![\varphi^{\#}]\!]$. (ii) For any order-$(n+1)$ closed disjunctive $\mu HFL(Z)$ formula φ, $\varphi^{\flat}$ is an order-n closed $\mu HFL(Z)$ formula such that $[\![\varphi]\!] = [\![\varphi^{\flat}]\!]$.*

Due to the connection between reachability problems and $\mu HFL(Z)$ validity checking problems discussed in Sect. 2.2, the theorem above implies that any order-n reachability game can be converted in polynomial time to order-$(n+1)$ may-reachability problem, and vice versa. Applications of this result are discussed in Sect. 5.

3 From Order-n Reachability Games to Order-$(n+1)$ May-Reachability

In this section, we show the translation $(\cdot)^{\#}$ from order-n $\mu HFL(Z)$ formulas to order-$(n+1)$ disjunctive $\mu HFL(Z)$ formulas. The idea is to transform each proposition φ (i.e. a formula of type $\star$) to a predicate $\varphi^{\#'}$ of type $\star \to \star$, so that true and false are respectively converted to the identity function $\lambda x.x$ and the constant function $\lambda x.\mathtt{false}$. We can then encode the conjunction $\varphi_1 \wedge \varphi_2$ as $\lambda x^{\star}.\varphi_1^{\#'}(\varphi_2^{\#'} x)$, which is equivalent to the identity function just if both $\varphi_1^{\#'}$ and $\varphi_2^{\#'}$ are.

The translation $(\cdot)^{\#}$ for formulas and types is defined as follows.

$$\varphi^{\#} = \varphi^{\#'}\ \mathtt{true} \qquad (e_1 \leq e_2)^{\#'} = \lambda x^{\star}.(e_1 \leq e_2 \wedge x) \qquad (\lambda x^{\kappa}.M)^{\#'} = \lambda x^{\kappa^{\#}}.M^{\#'}$$

$$(\varphi_1 \varphi_2)^{\#'} = \varphi_1^{\#'}\varphi_2^{\#'} \qquad (\varphi\, e)^{\#'} = \varphi^{\#'}\, e \qquad (\mu x^{\tau}.\varphi)^{\#'} = \mu x^{\tau^{\#}}.\varphi^{\#'}$$

$$(\varphi_1 \vee \varphi_2)^{\#'} = \lambda x^{\star}.\varphi_1^{\#'} x \vee \varphi_2^{\#'} x \qquad (\varphi_1 \wedge \varphi_2)^{\#'} = \lambda x^{\star}.\varphi_1^{\#'}(\varphi_2^{\#'} x)$$

$$\mathtt{Int}^{\#} = \mathtt{Int} \qquad \star^{\#} = \star \to \star \qquad (\kappa \to \tau)^{\#} = \kappa^{\#} \to \tau^{\#}.$$

Example 3. Consider the formula $\varphi := (\mu p^{\mathtt{Int} \to \star}.\lambda y.y = 0 \vee (p\,(y-1) \wedge p\,(y+1)))\, n$ (where n is an integer constant). The translation (followed by β-reductions for simplification) yields:

$$(\mu p^{\mathtt{Int} \to \star \to \star}.\lambda y.\lambda x^{\star}.(y = 0 \wedge x) \vee p\,(y-1)\,(p\,(y+1)\,x))\, n\ \mathtt{true}.$$

The following theorem states the correctness of the translation. The proof is given in the full version [1].

Theorem 3. *If φ is an order-n closed $\mu HFL(Z)$ formula, then $\varphi^{\#}$ is an order-$(n+1)$ closed disjunctive $\mu HFL(Z)$ formula, and $[\![\varphi]\!] = [\![\varphi^{\#}]\!]$.*

4   From Order-$(n+1)$ May-Reachability to Order-n Reachability Games

In this section, we show the translation $(\cdot)^\flat$ from order-$(n + 1)$ disjunctive μHFL(Z) formulas to order-n μHFL(Z) formulas. The translation $(\cdot)^\flat$ is much more involved than the translation $(\cdot)^\#$.

To see how such translation can be achieved, let us recall the formula $\varphi_{ex1} := \varphi_{sum}\, n\, (\lambda r.r < n)$ in Example 2, where $\varphi_{sum} : \mathtt{Int} \to (\mathtt{Int} \to \star) \to \star$ is:

$$\mu sum.\lambda x.\lambda k.x < 0 \vee (x = 0 \wedge k\, 0) \vee (x > 0 \wedge sum\, (x-1)\, (\lambda y.k(x+y))).$$

Note that the order of the formula above is 1. We wish to construct a formula ψ of order 0, such that $[\![\varphi_{ex1}]\!] = [\![\psi]\!]$. Recall that, by Fact 1, $[\![\varphi_{ex1}]\!] = \top$ just if $\varphi_{ex1} \longrightarrow^* \mathtt{true}$. There are two cases where the formula φ_{ex1} may be reduced to $\mathtt{true}$: (i) φ_{ex1} is reduced to $\mathtt{true}$ without the order-0 argument $\lambda r.r < n$ being called; and (ii) φ_{ex1} is reduced to $(\lambda r.r < n)m$ for some m, and then $(\lambda r.r < n)m$ is reduced to $\mathtt{true}$. Let $\varphi_{sum_0}\, n$ be the condition for the first case to occur, and let $\varphi_{sum_1}\, n\, m$ be the condition that φ_{ex1} is reduced to $(\lambda r.r < n)m$. Then, φ_{sum_0} and φ_{sum_1} can be expressed as follows.

$$\varphi_{sum_0} := \mu sum_0.\lambda x.x < 0 \vee (x > 0 \wedge sum_0\, (x-1)).$$
$$\varphi_{sum_1} := \mu sum_1.\lambda x.\lambda z.(x = 0 \wedge z = 0) \vee (x > 0 \wedge \exists y.sum_1\, (x-1)\, y \wedge z = x + y).$$

To understand the formula φ_{sum_1}, notice that $\varphi_{sum}\, (x-1)\, (\lambda y.k(x+y))$ is reduced to $k\, z$ just if $sum\, (x-1)\, (\lambda y.k(x+y))$ is first reduced to $(\lambda y.k(x+y))y$ for some y (the condition for which is expressed by $sum_1\, (x-1)\, y$), and $z = x + y$ holds.

Using φ_{sum_0} and φ_{sum_1} above, the formula φ_{ex1} can be translated to the order-0 formula $\varphi_{sum_0}\, n \vee \exists r.\varphi_{sum_1}\, n\, r \wedge r < n$. In general, if φ is an order-1 (disjunctive) formula of type $\mathtt{Int}^k \to (\mathtt{Int}^{\ell_1} \to \star) \to \cdots \to (\mathtt{Int}^{\ell_m} \to \star) \to \star$ and $\psi_i\ (i \in \{1,\ldots,m\})$ is a formula of type $\mathtt{Int}^{\ell_i} \to \star$, then $\varphi\, \widetilde{e}_{1,\ldots,k}\, \psi_1 \cdots \psi_m$ can be translated to an order-0 formula of the form:

$$\varphi_0\, \widetilde{e}_{1,\ldots,k} \vee \bigvee_{i \in \{1,\ldots,m\}} \exists \widetilde{y}_{1,\ldots,\ell_i}.(\varphi_i\, \widetilde{e}_{1,\ldots,k}\, \widetilde{y}_{1,\ldots,\ell_i} \wedge \psi_i\, \widetilde{y}_{1,\ldots,\ell_i}),$$

where the part $\varphi_0\, \widetilde{e}_{1,\ldots,k}$ expresses the condition for $\varphi\, \widetilde{e}_{1,\ldots,k}\, \psi_1 \cdots \psi_m$ to be reduced to $\mathtt{true}$ without ψ_i being called, and the part $\varphi_i\, \widetilde{e}_{1,\ldots,k}\, \widetilde{y}_{1,\ldots,\ell_i}$ expresses the condition for $\varphi\, \widetilde{e}_{1,\ldots,k}\, \psi_1 \cdots \psi_m$ to be reduced to $\psi_i\, \widetilde{y}_{1,\ldots,\ell_i}$.

For higher-order formulas, the translation is more involved. To simplify the formalization, we assume that a formula as an input or output of our translation is given in the form (Θ, D, φ_0), called an *equation system*; here D is a set of mutually recursive fixpoint equations of the form $\{F_1\, \widetilde{x}_1 =_\mu \varphi_1, \ldots, F_n\, \widetilde{x}_n =_\mu \varphi_n\}$ and Θ is the type environment for $F_1, \ldots, F_n$. We sometimes omit Θ and just write (D, φ_0). Here, each $\varphi_i\ (i \in \{0,\ldots,n\})$ should be fixpoint-free, φ_0 is well-typed under Θ, and $\varphi_i\ (i \in \{1,\ldots,n\})$ should have some type τ_i under the type environment $\Theta, x_{i,1}{:}\kappa_{i,1}, \ldots, x_{i,m_i}{:}\kappa_{i,m_i}$, where $\Theta(F_i) = \kappa_{i,1} \to \cdots \to \kappa_{i,m_i} \to \tau_i$

and $\widetilde{x}_i = x_{i,1} \cdots x_{i,m_i}$. The μHFL(Z) formula $(D, \varphi_0)^\mu$ represented by (Θ, D, φ_0) is defined by:

$$(\emptyset, \varphi)^\mu = \varphi \quad (D \cup \{F\,\widetilde{x} =_\mu \psi\}, \varphi)^\mu = ([\mu F.\lambda\widetilde{x}.\psi/F]D, [\mu F.\lambda\widetilde{x}.\psi/F]\varphi)^\mu.$$

We write $[\![(D, \varphi)]\!]$ for $[\![(D, \varphi)^\mu]\!]$.

For an equation system as an input of our translation, we further assume, without loss of generality, the following conditions.

(I) Each φ_i ($i \in \{1, \ldots, n\}$) on the right-hand side of a definition in D has type $\star$ and is generated by the following grammar (where the metavariable x may be a fixpoint variable F_j or its parameters):

$$\varphi ::= x \mid \varphi_1 \vee \varphi_2 \mid e_1 \le e_2 \wedge \varphi \mid \varphi_1\varphi_2 \mid \varphi\,e. \tag{1}$$

In particular, (i) φ_i is a disjunctive μHFL(Z) formula, (ii) φ_i contains neither λ-abstractions nor fixpoint operators, and (iii) a formula of the form $e_1 \le e_2$ may occur only in the form $e_1 \le e_2 \wedge \varphi$.

(II) Every integer predicate (i.e., a formula of type of the form $\mathtt{Int}^\ell \to \star$ with $\ell \ge 0$) that occurs in an argument position has the same arity M. In other words, in any function type $\kappa \to \tau$, either $\kappa = \mathtt{Int}^M \to \star$, or $\mathrm{ord}(\kappa) \ne 0$.

(III) The "main formula" φ_0 is a formula of the form $F\,\lambda\widetilde{x}_{1,\ldots,M}.\mathtt{true}$.

Note that the assumption above does not lose generality. Given an order-$(n+1)$ disjunctive μHFL(Z) formula φ, it can be first transformed to a formula of the form $\varphi'\,\mathtt{true}$, where $\mathtt{true}$ does not occur on the right-hand side of any conjunction in φ'. We then set M to the largest arity of integer predicates that occur in argument positions in $\varphi'\,\mathtt{true}$, and raise the arity of every integer predicate argument to M by adding dummy arguments. For example, given

$$(\lambda f^{\mathtt{Int}\to\star}.f\,1)((\lambda g^{\mathtt{Int}\to\mathtt{Int}\to\star}.g\,1)(\lambda x^{\mathtt{Int}}.\lambda y^{\mathtt{Int}}.x \le y)),$$

we can set M to 2, and replace the formula with:

$$(\lambda f'^{\mathtt{Int}\to\mathtt{Int}\to\star}.f'\,1\,0)\lambda z_1.\lambda z_2.((\lambda g^{\mathtt{Int}\to\mathtt{Int}\to\star}.g\,1)(\lambda x^{\mathtt{Int}}.\lambda y^{\mathtt{Int}}.x \le y))\,z_1.$$

Here, we have inserted dummy (actual and formal) parameters 0 and z_2 to increase the arities of f and the argument of $(\lambda f^{\mathtt{Int}\to\star}.f\,1)$. We can then apply λ-lifting to remove λ-abstractions and generate a set of top-level definitions D.

We translate each equation $F\,y_1 \cdots y_m =_\mu \varphi$ in D as follows. We first decompose the formal parameters $y_1, \ldots, y_m$ to two parts: $y_1, \ldots, y_j$ and $y_{j+1}, \ldots, y_m$, where the orders of (the types of) $y_{j+1}, \ldots, y_m$ are at most 0, and the order of y_j is at least 1; note that the sequences $y_1, \ldots, y_j$ and $y_{j+1}, \ldots, y_m$ are possibly empty. We further decompose $y_{j+1}, \ldots, y_m$ into order-0 variables $x_1, \ldots, x_k$ and integer variables $z_1, \ldots, z_p$ (thus, $j + k + p = m$). Formally, the decomposition of formal parameters is defined by $\mathtt{decomparg}(\epsilon, \star) = (\epsilon, \epsilon, \epsilon)$ and

$$\mathbf{decomparg}(u \cdot \widetilde{y}, \kappa \to \tau) =$$
$$\begin{cases} ((u:\kappa) \cdot \mathcal{K}, \widetilde{x}, \widetilde{z}) & \text{if } \mathbf{decomparg}(\widetilde{y}, \tau) = (\mathcal{K}, \widetilde{x}, \widetilde{z}), \mathcal{K} \neq \epsilon \\ (u:\kappa, \widetilde{x}, \widetilde{z}) & \text{if } \mathrm{ord}(\kappa) > 0, \mathbf{decomparg}(\widetilde{y}, \tau) = (\epsilon, \widetilde{x}, \widetilde{z}) \\ (\epsilon, u \cdot \widetilde{x}, \widetilde{z}) & \text{if } \kappa = \mathtt{Int}^M \to \star, \mathbf{decomparg}(\widetilde{y}, \tau) = (\epsilon, \widetilde{x}, \widetilde{z}) \\ (\epsilon, \widetilde{x}, u \cdot \widetilde{z}) & \text{if } \kappa = \mathtt{Int}, \mathbf{decomparg}(\widetilde{y}, \tau) = (\epsilon, \widetilde{x}, \widetilde{z}) \end{cases}$$

Here, $\mathbf{decomparg}(\widetilde{y}_{1,\dots,m}, \Theta(F))$ decomposes the sequence of variables $\widetilde{y}_{1,\dots,m}$ and returns a triple $(\mathcal{K}, \widetilde{x}, \widetilde{z})$, where $\mathcal{K}$ is the type environment for $y_1, \dots, y_j$, $\widetilde{x}$ is the sequence of integer predicate variables, and $\widetilde{z}$ is the sequence of integer variables.

For example, given an equation $F u_1 u_2 u_3 u_4 u_5 =_\mu \varphi$, where $\Theta(F) = \mathtt{Int} \to ((\mathtt{Int} \to \star) \to \star) \to \mathtt{Int} \to (\mathtt{Int} \to \star) \to \mathtt{Int} \to \star$, the formal parameters $u_1 \cdots u_5$ are decomposed as follows.

$$\mathbf{decomparg}(u_1 \cdots u_5, \Theta(F)) = (\{u_1 : \mathtt{Int}, u_2 : (\mathtt{Int} \to \star) \to \star\}, u_4, u_3 u_5).$$

Given an equation $F \widetilde{y} =_\mu \varphi$ where $\mathbf{decomparg}(\widetilde{y}, \Theta(F)) = (\mathcal{K}, \widetilde{x}_{1,\dots,k}, \widetilde{z})$ with $\mathcal{K} = y_1{:}\kappa_1, \dots, y_j{:}\kappa_j$, we generate equations for new fixpoint variables $F_0, \dots, F_k$. As in the order-1 case, for $i \in \{1, \dots, k\}$, $F_i \widetilde{\varphi}'_{1,\dots,j} \widetilde{z} \widetilde{u}_{1,\dots,M}$ represents the condition for $F \widetilde{\varphi}_{1,\dots,j}$ to be reduced to $x_i \widetilde{u}_{1,\dots,M}$ (where $\widetilde{\varphi}'_{1,\dots,j}$ is the sequence of formulas obtained by translating $\widetilde{\varphi}_{1,\dots,j}$ in a recursive manner). F_0 is a new component required to deal with higher-order formulas; it is used to compute the condition for $F \widetilde{y}$ to be reduced to $x \widetilde{u}_{1,\dots,\ell_i}$ for some order-0 predicate x, which has been passed through higher-order parameters $\widetilde{y}_{1,\dots,j}$. For example, consider a formula $F (G x) y$ where $F : ((\mathtt{Int} \to \star) \to \star) \to (\mathtt{Int} \to \star) \to \star, G : (\mathtt{Int} \to \star) \to (\mathtt{Int} \to \star) \to \star$. Then, the condition for $F (G x) y$ to be reduced to $y n$ is computed by using F_1, while the condition for $F (G x) y$ to be reduced to $x n$ is computed by using F_0; see Example 4 for a concrete version of this example.

To compute $F_0, \dots, F_k$, we translate each subformula φ of the body of F to: $(\varphi_*, \varphi_0, \varphi_1, \dots, \varphi_k, \varphi_{k+1}, \dots, \varphi_{k+\mathbf{gar}(\tau)})$, where τ is the type of φ, and $\mathbf{gar}(\tau)$ denotes the number of order-0 arguments passed after the last argument of order greater than 0. More precisely, we define the decomposition of types as follows.

$$\mathbf{decomp}(\star) = (\epsilon, \epsilon, 0)$$

$$\mathbf{decomp}(\kappa \to \tau) = \begin{cases} (\kappa \cdot \widetilde{\kappa}, m, n) & \text{if } \mathbf{decomp}(\tau) = (\widetilde{\kappa}, m, n), \widetilde{\kappa} \neq \epsilon \\ (\kappa, m, n) & \text{if } \mathrm{ord}(\kappa) > 0, \mathbf{decomp}(\tau) = (\epsilon, m, n) \\ (\epsilon, m+1, n) & \text{if } \kappa = \mathtt{Int}^M \to \star, \mathbf{decomp}(\tau) = (\epsilon, m, n) \\ (\epsilon, m, n+1) & \text{if } \kappa = \mathtt{Int}, \mathbf{decomp}(\tau) = (\epsilon, m, n) \end{cases}$$

Then, $\mathbf{gar}(\tau)$ denotes m when $\mathbf{decomp}(\tau) = (\widetilde{\kappa}, m, n)$. For example, for $\tau = (\mathtt{Int} \to \star) \to ((\mathtt{Int} \to \star) \to \star) \to (\mathtt{Int} \to \star) \to \mathtt{Int} \to (\mathtt{Int} \to \star) \to \star$, $\mathbf{decomp}(\tau) = ((\mathtt{Int} \to \star) \cdot ((\mathtt{Int} \to \star) \to \star), 2, 1)$; hence $\mathbf{gar}(\tau) = 2$. Here, $\varphi_1, \dots, \varphi_k$ are analogous to $F_1, \dots, F_k$: they are used for computing the condition for $\varphi \widetilde{\psi}$ to be reduced to $x_i \widetilde{n}$. Similarly, φ_{k+i} (where $i \in \{1, \dots, \mathbf{gar}(\tau)\}$) is used for computing the condition for $\varphi \widetilde{\psi}$ to be reduced to $\psi_i \widetilde{n}$, where ψ_i is the i-th order-0 argument of φ. The component φ_0 is analogous to F_0, and used to compute the condition for $\varphi \widetilde{\psi}$ to be reduced to $x \widetilde{n}$, where x is an order-0 predicate passed through higher-order arguments of φ. The other component φ_* is similar to φ_0, but the target predicate x may have already been set inside φ_*.

Based on the intuition above, we formalize the translation of a formula as the relation: $\mathcal{K}; \widetilde{x}_{1,\ldots,k} \vdash_\Theta \varphi : \tau \rightsquigarrow (\varphi_*, \varphi_0, \ldots, \varphi_{k+\mathtt{gar}(\tau)})$. Here, Θ denotes the type environment for fixpoint variables defined by D. If φ is a subformula of the body of F, and F is defined by $F\widetilde{y} =_\mu \varphi_F$, then $\mathcal{K}$ and $\widetilde{x}_{1,\ldots,k}$ are set to $\mathcal{K}_F, \widetilde{z} : \overline{\mathtt{Int}}$ and $\widetilde{x}_F$ respectively, where $\mathtt{decomparg}(\widetilde{y}, \Theta(F)) = (\mathcal{K}_F, \widetilde{x}_F, \widetilde{z})$.

The translation rules are given in Fig. 2. We explain the main rules below. In the rule TR-VARG for an order-0 variable x_i (which should disappear after the translation), $\varphi_j \, \widetilde{z}_{1,\ldots,M} \, \widetilde{w}_{1,\ldots,M}$ should represent the condition for $x_i \, \widetilde{z}_{1,\ldots,M} \longrightarrow^*$ $x_j \, \widetilde{w}_{1,\ldots,M}$; thus φ_j is defined so that $\widetilde{z}_{1,\ldots,M} \, \widetilde{w}_{1,\ldots,M}$ is equivalent to $\mathtt{true}$ just if $i = j$ and $\widetilde{z}_{1,\ldots,M} = \widetilde{w}_{1,\ldots,M}$. In the rule TR-VAR for a variable y in $\mathcal{K}$, the output of the translation is constructed from $(y_*, y_0, y_1, \ldots, y_m)$, whose values will be provided by the environment. Because the environment does not know order-0 variables $x_1, \ldots, x_k$, we use y_0 to compute the condition for $y \, \widetilde{\psi}$ to be reduced to $x_i \, \widetilde{m}$. The rule TR-VARF for fixpoint variables is almost the same as TR-VAR, except that the component F_0 is reused for F_*. The rationale for this is as follows: both φ_* and φ_0 are used for computing the condition for a target order-0 predicate variable (which is set by the environment) to be reached, and the only difference between them is that the target predicate may have already been set in φ_*, but since F is a closed formula, such distinction does not make any difference; hence F_0 and F_* need not be distinguished from each other.

In the rule TR-APP, the first two components ($\varphi_*(\psi_*, \ldots)$ and $\varphi_0(\psi_0, \ldots)$) are used for computing the condition for some target predicates (set by the environment) to be reached, and the next k components ($\varphi_1(\psi_1, \ldots), \ldots, \varphi_k(\psi_k, \ldots)$) are used for computing the condition for predicate $x_1, \ldots, x_k$ to be reached. The rule TR-APPG is another rule for applications, where the argument ψ is an order-0 predicate. The component ξ_j of the output is used for computing the condition for the predicate x_i to be reached (i.e., the condition for a formula of the form $\varphi \psi \widetilde{\psi}'$ to be reduced to $x_i \, \widetilde{w}_{1,\ldots,\ell_j}$, where $\widetilde{\psi}'$ consists of order-0 predicates and integer arguments $\widetilde{z}_{1,\ldots,p}$). The formula $\varphi \psi \widetilde{\psi}'$ may be reduced to $x_i \, \widetilde{w}_{1,\ldots,\ell_j}$ if either (i) $\varphi \psi \widetilde{\psi}' \longrightarrow^* x_i \, \widetilde{w}_{1,\ldots,\ell_j}$ without ψ being called, or (ii) $\varphi \psi \widetilde{\psi}'$ is reduced to $\psi \widetilde{z} \widetilde{u}$ for some $\widetilde{u}$, and $\psi \widetilde{z} \widetilde{u}$ is reduced to $x_i \, \widetilde{w}_{1,\ldots,\ell_j}$. The part $\varphi_j \widetilde{z} \widetilde{w}$ represents the former condition, and the part $\exists \widetilde{u}. \cdots$ represents the latter.

Example 4. Consider $S(\lambda x.\mathtt{true})$, where S is defined by:

$$S\,t =_\mu F(G\,t)\,t \qquad F\,v\,w =_\mu v\,H \vee w\,2 \qquad G\,p\,q =_\mu p\,1 \qquad H\,x =_\mu H\,x.$$

There are the following two ways for $S\,t$ to be reduced to $t\,n$ for some n.

$$S\,t \longrightarrow F(G\,t)\,t \longrightarrow G\,t\,H \vee t\,2 \longrightarrow G\,t\,H \longrightarrow t\,1$$
$$S\,t \longrightarrow F(G\,t)\,t \longrightarrow G\,t\,H \vee t\,2 \longrightarrow t\,2.$$

$$\varphi_j = \begin{cases} \lambda \widetilde{z}_{1,\ldots,M}.\lambda \widetilde{w}_{1,\ldots,M}.\wedge_{p=1,\ldots,M}(z_p = w_p), & \text{if } j = i \\ \lambda \widetilde{z}_{1,\ldots,M}.\lambda \widetilde{w}_{1,\ldots,M}.\mathtt{false} & \text{otherwise} \end{cases}$$

$$\overline{\mathcal{K}; \widetilde{x}_{1,\ldots,k} \vdash_\Theta x_i : \mathtt{Int}^M \to \star \rightsquigarrow (\varphi_*, \varphi_0, \ldots, \varphi_k)} \qquad \text{(Tr-VarG)}$$

$$\frac{\mathtt{decomp}(\mathcal{K}(y)) = (\widetilde{\kappa}, m, p)}{\mathcal{K}; \widetilde{x}_{1,\ldots,k} \vdash_\Theta y : \mathcal{K}(y) \rightsquigarrow (y_*, \underbrace{y_0, \ldots, y_0}_{k+1}, y_1, \ldots, y_m)} \qquad \text{(Tr-Var)}$$

$$\frac{\mathtt{decomp}(\Theta(F)) = (\widetilde{\kappa}, m, p)}{\mathcal{K}; \widetilde{x}_{1,\ldots,k} \vdash_\Theta F : \Theta(F) \rightsquigarrow (F_0, \underbrace{F_0, \ldots, F_0}_{k+1}, F_1, \ldots, F_m)} \qquad \text{(Tr-VarF)}$$

$$\frac{\mathcal{K}; \widetilde{x}_{1,\ldots,k} \vdash_\Theta \varphi : \star \rightsquigarrow (\varphi_*, \varphi_0, \ldots, \varphi_k) \qquad \psi_j = \lambda \widetilde{z}_{1,\ldots,M}.e_1 \le e_2 \wedge \varphi_j \, \widetilde{z}_{1,\ldots,M}}{\mathcal{K}; \widetilde{x}_{1,\ldots,k} \vdash_\Theta e_1 \le e_2 \wedge \varphi : \star \rightsquigarrow (\psi_*, \psi_0, \ldots, \psi_k)} \qquad \text{(Tr-Le)}$$

$$\mathtt{ord}(\kappa_0 \to \tau) > 1 \qquad \mathtt{gar}(\kappa_0 \to \tau) = m \qquad \mathtt{gar}(\kappa_0) = m'$$
$$\mathcal{K}; \widetilde{x}_{1,\ldots,k} \vdash_\Theta \varphi : \kappa_0 \to \tau \rightsquigarrow (\varphi_*, \varphi_0, \ldots, \varphi_{k+m})$$
$$\frac{\mathcal{K}; \widetilde{x}_{1,\ldots,k} \vdash_\Theta \psi : \kappa_0 \rightsquigarrow (\psi_*, \psi_0, \ldots, \psi_{k+m'})}{\begin{aligned} \mathcal{K}; \widetilde{x}_{1,\ldots,k} \vdash_\Theta \varphi\psi : \tau \rightsquigarrow (&\varphi_*(\psi_*, \psi_0, \psi_{k+1}, \ldots, \psi_{k+m'}), \varphi_0(\psi_0, \psi_0, \psi_{k+1}, \ldots, \psi_{k+m'}), \\ &\varphi_1(\psi_1, \psi_0, \psi_{k+1}, \ldots, \psi_{k+m'}), \ldots, \varphi_k(\psi_k, \psi_0, \psi_{k+1}, \ldots, \psi_{k+m'}), \\ &\varphi_{k+1}(\psi_0, \psi_{k+1}, \ldots, \psi_{k+m'}), \ldots, \varphi_{k+m}(\psi_0, \psi_{k+1}, \ldots, \psi_{k+m'})) \end{aligned}} \qquad \text{(Tr-App)}$$

$$\mathtt{decomp}(\tau) = (\epsilon, m-1, p)$$
$$\mathcal{K}; \widetilde{x}_{1,\ldots,k} \vdash_\Theta \varphi : (\mathtt{Int}^M \to \star) \to \tau \rightsquigarrow (\varphi_*, \varphi_0, \ldots, \varphi_{k+m})$$
$$\mathcal{K}; \widetilde{x}_{1,\ldots,k} \vdash_\Theta \psi : \mathtt{Int}^M \to \star \rightsquigarrow (\psi_*, \psi_0, \ldots, \psi_k)$$
$$\frac{\xi_j = \lambda \widetilde{z}_{1,\ldots,p}.\lambda \widetilde{w}_{1,\ldots,M}.\varphi_j \, \widetilde{z} \, \widetilde{w} \vee \exists \widetilde{u}_{1,\ldots,M}.(\varphi_{k+1} \, \widetilde{z} \, \widetilde{u}_{1,\ldots,M} \wedge \psi_j \, \widetilde{u}_{1,\ldots,M} \, \widetilde{w}_{1,\ldots,M})}{\mathcal{K}; \widetilde{x}_{1,\ldots,k} \vdash_\Theta \varphi\psi : \tau \rightsquigarrow (\xi_*, \xi_0, \ldots, \xi_k, \varphi_{k+2}, \ldots, \varphi_{k+m})} \qquad \text{(Tr-AppG)}$$

$$\frac{\mathcal{K}; \widetilde{x}_{1,\ldots,k} \vdash_\Theta \varphi : \mathtt{Int} \to \tau \rightsquigarrow (\varphi_*, \varphi_0, \ldots, \varphi_{\ell+k})}{\mathcal{K}; \widetilde{x}_{1,\ldots,k} \vdash_\Theta \varphi e : \tau \rightsquigarrow (\varphi_* \, e, \varphi_0 \, e, \ldots, \varphi_{\ell+k} \, e)} \qquad \text{(Tr-AppI)}$$

$$\mathcal{K}; \widetilde{x}_{1,\ldots,k} \vdash_\Theta \varphi : \star \rightsquigarrow (\varphi_*, \varphi_0, \ldots, \varphi_k) \qquad \mathcal{K}; \widetilde{x}_{1,\ldots,k} \vdash_\Theta \psi : \star \rightsquigarrow (\psi_*, \psi_0, \ldots, \psi_k)$$
$$\frac{\xi_j = \lambda \widetilde{z}_{1,\ldots,M}.\varphi_j \, \widetilde{z}_{1,\ldots,M} \vee \psi_j \, \widetilde{z}_{1,\ldots,M}}{\mathcal{K}; \widetilde{x}_{1,\ldots,k} \vdash_\Theta \varphi \vee \psi : \star \rightsquigarrow (\xi_*, \xi_0, \ldots, \xi_k)} \qquad \text{(Tr-Disj)}$$

$$\mathtt{decomparg}(\widetilde{w}, \Theta(F)) = (\widetilde{y} : \widetilde{\kappa}, \widetilde{x}_{1,\ldots,k}, \widetilde{z})$$
$$y_1 : \kappa_1, \ldots, y_m : \kappa_m, \widetilde{z} : \widetilde{\mathtt{Int}}; \widetilde{x}_{1,\ldots,k} \vdash_\Theta \varphi : \star \rightsquigarrow (\varphi_*, \varphi_0, \ldots, \varphi_k)$$
$$\begin{cases} \widetilde{y}_i = (y_{i,*}, y_{i,0}, \ldots, y_{i,\mathtt{gar}(\kappa_i)}) & \widetilde{y'_i} = (y_{i,0}, \ldots, y_{i,\mathtt{gar}(\kappa_i)}) & \text{if } i \in \{1, \ldots, m\}, \, \kappa_i \ne \mathtt{Int} \\ \widetilde{y}_i = y_i & \widetilde{y'_i} = y_i & \text{if } i \in \{1, \ldots, m\}, \, \kappa_i = \mathtt{Int} \end{cases}$$
$$\frac{}{\vdash_\Theta (F \, \widetilde{w} =_\mu \varphi) \rightsquigarrow \{F_0 \, \widetilde{y}_1 \cdots \widetilde{y}_m \, \widetilde{z} =_\mu \varphi_*\} \cup \{F_i \, \widetilde{y'_1} \cdots \widetilde{y'_m} \, \widetilde{z} =_\mu \varphi_i \mid i \in \{1, \ldots, k\}\}} \qquad \text{(Tr-Def)}$$

$$\frac{D' = \bigcup \{D'' \mid \vdash_\Theta (F \, \widetilde{y} =_\mu \varphi) \rightsquigarrow D'' \mid F \, \widetilde{y} =_\mu \varphi \in D\}}{(D, S \, \lambda \widetilde{z}.\mathtt{true}) \rightsquigarrow (D', \exists \widetilde{z}.S_1 \, \widetilde{z})} \qquad \text{(Tr-Main)}$$

Fig. 2. Translation from order-$(n+1)$ disjunctive μHFL(Z) to morder-n μHFL(Z).

The output of our transformations (with some simplification) is $\exists z.S_1 \, z$ where:

$$S_1 =_\mu \lambda w_1.F_0 \, (\lambda w_1.G_0 \, w_1 \vee G_1 \, w_1, G_0, G_2) \, w_1 \vee F_1 \, (G_0, G_2) \, w_1$$

$$F_0 \, (v_*, v_0, v_1) =_\mu \lambda z_1.v_* \, z_1 \vee \exists u_1.v_1 \, u_1 \wedge H_0 \, u_1 \, z_1$$
$$F_1 \, (v_0, v_1) =_\mu \lambda z_1.v_0 \, z_1 \vee (\exists u_1.v_1 \, u_1 \wedge H_0 \, u_1 \, z_1) \vee 2 = z_1$$
$$G_0 =_\mu \lambda w_1.\mathtt{false} \quad G_1 =_\mu \lambda w_1.1 = w_1 \quad G_2 =_\mu \lambda w_1.\mathtt{false} \quad H_0 \, x =_\mu H_0 \, x.$$

Notice that the formula $S_1 \, z$ has the following two reduction sequences that lead to the conditions of the form $z = n$ for some n.

$$S_1 \, z \longrightarrow^* F_0 \, (\lambda w_1.G_0 \, w_1 \vee G_1 \, w_1, G_0, G_2) \, z \longrightarrow^* (\lambda w_1.G_0 \, w_1 \vee G_1 \, w_1)z \longrightarrow^* 1 = z$$
$$S_1 \, z \longrightarrow^* F_1 \, (G_0, G_2) \, z \longrightarrow^* G_0 \, z \vee (\exists u_1.G_2 \, u_1 \wedge H_0 \, u_1 \, z) \vee 2 = z \longrightarrow^* 2 = z.$$

The former reduction sequence corresponds to the reduction sequence of the original formula $S \, t \longrightarrow^* t \, 1$ where t embedded in the first argument of F (in $F \, (G \, t) \, t$) is called, and the latter reduction sequence corresponds to the reduction sequence $S \, t \longrightarrow^* t \, 2$ where the second argument t of F (in $F \, (G \, t) \, t$) is called. Note that the first condition $1 = z$ has been computed by using F_0, and the second condition $2 = z$ has been computed by using F_1.   □

The following theorem states the correctness of the translation. A proof of the theorem and more examples of the translation are given in [1].

Theorem 4. If $(D, S \, \lambda \widetilde{z}_{1,\dots,M}.\mathtt{true}) \rightsquigarrow (D', \psi)$, then $[\![(D, S \, \lambda \widetilde{z}_{1,\dots,M}.\mathtt{true})]\!] = [\![(D', \psi)]\!]$.

5   Applications

As mentioned already, the translation from order-n reachability games to order-$(n{+}1)$ may-reachability enables us to use automated (un)reachability checkers for solving the reachability game problem, and the translation in the other direction enables us to use, for example, reachability game solvers for non-higher-order programs as a may-reachability checker for order-1 programs.

As a direct application of the former translation, we have applied it to the νHFL(Z) solver RETHFL [11], which is a refinement-type-based validity checker for formulas of νHFL(Z), the fragment of HFL(Z) without least fix-point operators (but with greatest fixpoint operators). The fragment νHFL(Z) is dual to μHFL(Z), in the sense that, for every closed formula φ of type $\star$ of μHFL(Z), there exists a νHFL(Z) formula $\overline{\varphi}$ such that φ is valid if and only if $\overline{\varphi}$ is invalid, and vice versa; $\overline{\varphi}$ is obtained from φ by just replacing each logical operator (including fixpoint operators) with its de Morgan dual, and $e_1 \leq e_2$ with $e_1 > e_2$. Using a refinement type system, RETHFL reduces the validity of a given νHFL(Z) formula in a sound (but incomplete) manner to an extended CHC (constraint Horn clauses) problem, where disjunction is allowed in the head of each clause, and passes the problem to an extended CHC solver called PCSat [21]. For a fragment of νHFL(Z) corresponding to *disjunctive* μHFL(Z), however, the reduced problem is actually an ordinary CHC problem, for which more efficient tools [5,9,17] can be invoked. Thus, we can use the translation in Sect. 3 to improve the efficiency of RETHFL.

From the benchmark suite of RETHFL [11] (which originates from [10]), we picked the "non-termination" benchmark set, which consists of formulas obtained from non-termination verification of higher-order programs. All the formulas in that benchmark set do not belong to (the dual of) disjunctive μHFL(Z) (in contrast, the problems in the other benchmark sets belong to disjunctive μHFL(Z), hence our translation is not required). We have implemented the translation in Sect. 3, applied it to the problems in the "non-termination" benchmark set, and then ran RETHFL with a CHC solver HoICE [5,6] as the back-end solver. We have compared the result with plain RETHFL (without the transformation), which uses the extended CHC solver PCSat.

Table 1. Experimental results. Times are in seconds, with the timeout of 180 s.

Input	RETHFL	RETHFL+i.s.	RETHFL+ tr.
fixpoint_nonterm	11.579	0.054	0.102
unfoldr_nonterm	timeout	unknown	4.22
indirect_e	16.832	0.035	0.066
alternate	unknown	unknown	unknown
fib_CPS_nonterm	timeout	0.047	0.075
foldr_nonterm	8.447	unknown	0.122
passing_cond	116.423	unknown	0.444
indirectHO_e	11.582	0.044	0.073
inf_closure	timeout	20.171	9.080
loopHO	timeout	0.026	0.121

The results are summarized in Table 1. The column 'RETHFL' shows the result of plain RETHFL with PCSat as the back-end extended CHC solver (since ordinary CHC solvers are inapplicable to this benchmark set, as explained above). The column 'RETHFL+i.s.' show the result of RETHFL where the subtyping relation has been replaced by the imprecise one (equivalent to that of Horus [4], a HoCHC solver that can also be viewed as a νHFL(Z) solver) so that the type checking problem is reduced to ordinary CHC solving. The column 'RETHFL+tr.' shows the result of RETHFL with our translation. In both 'RETHFL+i.s.' and 'RETHFL+tr.', HoICE was used as the back-end CHC solver. The entry "unknown" indicates that the solver terminated with the answer "ill-typed", in which case, we do not know whether the formula is valid or invalid, due to the incompleteness of the underlying refinement type system.[2] The refinement type system used in 'RETHFL+i.s.' is less precise than the one used in RETHFL; hence, it returns more unknowns. As clear from the table, our translation significantly improved the efficiency of RETHFL.

---

[2] Although the understanding of the refinement type systems RETHFL is not required below, interested readers may wish to consult [11].

The translation in the other direction given in Sect. 4 also helps RETHFL, especially for relaxing the limitation caused by the incompleteness of the underlying refinement type system. For example, consider the formula S true, where:

$$S\,t =_{\mu} App\,(\lambda x.x \neq 0 \wedge t)\,0 \qquad App\,p\,y =_{\mu} p\,y \vee App\,(\lambda z.p(z-1))\,(y+1).$$

The formula is invalid, but RETHFL (nor Horus [4], a higher-order CHC solver based on a refinement type system) cannot prove the validity of the dual formula, due to the incompleteness of the refinement type system. The translation in Sect. 4 yields the following order-0 formula:[3]

$$S_1 =_{\mu} \exists x.App_1\,0\,x \wedge x \neq 0$$
$$App_1\,y\,z =_{\mu} y = z \vee \exists w.App_1\,(y+1)\,w \wedge w - 1 = z.$$

Here, $App_1\,y\,z$ intuitively means that $App\,p\,y$ can be reduced to $p\,z$. The underlying type system of RETHFL is complete for order-0 formulas, and indeed, the order-0 formula above can automatically be proved invalid by RETHFL.

6 Related Work

The relationship between order-n reachability games and order-$(n+1)$ may-reachability has some deep connection to the relationship between order-n tree languages and order-$(n+1)$ word languages [2,3,7], intuitively because the may-reachability problem is concerned about the set of "paths" of the execution tree of a given program, whereas the reachability game problem is also concerned about the branching structures of the execution tree. Indeed, our translations (especially, the use of φ_* and φ_0 components in the translation in Sect. 4) have been inspired by Asada and Kobayashi's translations between tree and word languages [3]. Kobayashi et al. [12] have also used a similar idea for a characterization of termination probabilities of higher-order probabilistic programs.

For finite-data programs (programs in Sect. 2.2 without integers), according to the complexity results on HORS model checking [14,18], both the order-n reachability game problem and the order-$(n+1)$ may-reachability game problem are n-EXPTIME complete, which imply that there are mutual translations between them. Concrete translations have, however, not been given (except unnatural translations through Turing machines). Also, the complexity-theoretic argument for the existence of translations does not apply in the presence of integers.

For HORS model checking, Parys [19] developed an order-decreasing transformation for higher-order grammars, which shares some ideas with our translation in Sect. 4. The details of the translations are however quite different. His translation makes use of finiteness in a crucial manner, and is not applicable in the presence of integers. Also, his translation is not size-preserving.

---

[3] We have implemented a prototype translator, but have not yet integrated it into RETHFL. For readability, here we show the formula obtained by some manual simplification of the automatically generated formula.

For order-1 programs, Kobayashi et al. [13] have shown that linear-time omega regular properties can be translated to order-0 HFL(Z) formulas. Our translation in Sect. 4 may be viewed as a higher-order extension of their translation, while the properties are restricted to may-reachability.

The fragment μHFL(Z) (or its dual fragment νHFL(Z)) is essentially (modulo the restriction of data domains to integers) equivalent to HoCHC [4], a higher-order extension of CHC. Therefore, the result of this paper should be useful also for improving HoCHC solvers.

7   Conclusion

We have shown translations between order-n reachability games and order-$(n+1)$ may-reachability, and proved their correctness. We have applied the translations to higher-order program verification, and obtained promising results in preliminary experiments. As mentioned in Sect. 6, our results are closely related to the correspondence between higher-order word and tree languages [3]. A deeper investigation of the relationship and generalization of the translations that subsume the related translations [3,12] are left for future work.

Acknowledgments. We would like to thank anonymous referees for useful comments. This work was supported by JSPS KAKENHI Grant Number JP20H05703 and JST SPRING Grant Number JPMJSP2108.

References

1. Asada, K., Katsura, H., Kobayashi, N.: On higher-order reachability games vs may reachability. CoRR abs/2203.08416 (2022). https://doi.org/10.48550/arXiv.2203.08416
2. Asada, K., Kobayashi, N.: On word and frontier languages of unsafe higher-order grammars. In: Chatzigiannakis, I., Mitzenmacher, M., Rabani, Y., Sangiorgi, D. (eds.) 43rd International Colloquium on Automata, Languages, and Programming, ICALP 2016, 11–15 July 2016, Rome, Italy. LIPIcs, vol. 55, pp. 111:1–111:13. Schloss Dagstuhl - Leibniz-Zentrum fuer Informatik (2016). https://doi.org/10.4230/LIPIcs.ICALP.2016.111
3. Asada, K., Kobayashi, N.: Size-preserving translations from order-(n+1) word grammars to order-n tree grammars. In: Ariola, Z.M. (ed.) 5th International Conference on Formal Structures for Computation and Deduction, FSCD 2020, June 29–July 6 2020, Paris, France (Virtual Conference). LIPIcs, vol. 167, pp. 22:1–22:22. Schloss Dagstuhl - Leibniz-Zentrum für Informatik (2020). https://doi.org/10.4230/LIPIcs.FSCD.2020.22
4. Burn, T.C., Ong, C.L., Ramsay, S.J.: Higher-order constrained Horn clauses for verification. Proc. ACM Program. Lang. **2**(POPL), 11:1–11:28 (2018). https://doi.org/10.1145/3158099
5. Champion, A., Chiba, T., Kobayashi, N., Sato, R.: ICE-based refinement type discovery for higher-order functional programs. J. Autom. Reason. **64**(7), 1393–1418 (2020). https://doi.org/10.1007/s10817-020-09571-y

6. Champion, A., Kobayashi, N., Sato, R.: HoIce: An ICE-based non-linear horn clause solver. In: Ryu, S. (ed.) Programming Languages and Systems - 16th Asian Symposium, APLAS 2018, Wellington, New Zealand, December 2–6, 2018, Proceedings. Lecture Notes in Computer Science, vol. 11275, pp. 146–156. Springer (2018). https://doi.org/10.1007/978-3-030-02768-1_8

7. Damm, W.: The IO- and OI-hierarchies. Theor. Comput. Sci. **20**, 95–207 (1982)

8. Grädel, E., Thomas, W., Wilke, T.: Automata, Logics, and Infinite Games: A Guide to Current Research, LNCS, vol. 2500. Springer, Cham (2002). https://doi.org/10.1007/3-540-36387-4

9. Hojjat, H., Rümmer, P.: The ELDARICA horn solver. In: 2018 Formal Methods in Computer Aided Design (FMCAD), pp. 1–7 (2018)

10. Iwayama, N., Kobayashi, N., Suzuki, R., Tsukada, T.: Predicate abstraction and CEGAR for νHFL$_Z$ validity checking. In: Pichardie, D., Sighireanu, M. (eds.) SAS 2020. LNCS, vol. 12389, pp. 134–155. Springer, Cham (2020). https://doi.org/10.1007/978-3-030-65474-0_7

11. Katsura, H., Iwayama, N., Kobayashi, N., Tsukada, T.: A new refinement type system for automated νHFL$_Z$ validity checking. In: Oliveira, B.C.S. (ed.) APLAS 2020. LNCS, vol. 12470, pp. 86–104. Springer, Cham (2020). https://doi.org/10.1007/978-3-030-64437-6_5

12. Kobayashi, N., Dal Lago, U., Grellois, C.: On the termination problem for probabilistic higher-order recursive programs. In: Proceedings of LICS 2019. IEEE (2019)

13. Kobayashi, N., Nishikawa, T., Igarashi, A., Unno, H.: Temporal verification of programs via first-order fixpoint logic. In: Chang, B.-Y.E. (ed.) SAS 2019. LNCS, vol. 11822, pp. 413–436. Springer, Cham (2019). https://doi.org/10.1007/978-3-030-32304-2_20

14. Kobayashi, N., Ong, C.-H.L.: Complexity of model checking recursion schemes for fragments of the modal mu-calculus. In: Albers, S., Marchetti-Spaccamela, A., Matias, Y., Nikoletseas, S., Thomas, W. (eds.) ICALP 2009. LNCS, vol. 5556, pp. 223–234. Springer, Heidelberg (2009). https://doi.org/10.1007/978-3-642-02930-1_19

15. Kobayashi, N., Sato, R., Unno, H.: Predicate abstraction and CEGAR for higher-order model checking. In: PLDI 2011, pp. 222–233. ACM Press (2011)

16. Kobayashi, N., Tsukada, T., Watanabe, K.: Higher-order program verification via HFL model checking. In: Ahmed, A. (ed.) ESOP 2018. LNCS, vol. 10801, pp. 711–738. Springer, Cham (2018). https://doi.org/10.1007/978-3-319-89884-1_25

17. Komuravelli, A., Gurfinkel, A., Chaki, S.: SMT-based model checking for recursive programs. Formal Methods Syst. Des. **48**(3), 175–205 (2016). https://doi.org/10.1007/s10703-016-0249-4

18. Ong, C.H.L.: On model-checking trees generated by higher-order recursion schemes. In: LICS 2006, pp. 81–90. IEEE Computer Society Press (2006)

19. Parys, P.: Higher-order model checking step by step. In: Bansal, N., Merelli, E., Worrell, J. (eds.) 48th International Colloquium on Automata, Languages, and Programming, ICALP 2021, July 12–16, 2021, Glasgow, Scotland (Virtual Conference). LIPIcs, vol. 198, pp. 140:1–140:16. Schloss Dagstuhl - Leibniz-Zentrum für Informatik (2021). https://doi.org/10.4230/LIPIcs.ICALP.2021.140

20. Rondon, P.M., Kawaguchi, M., Jhala, R.: Liquid types. In: PLDI 2008, pp. 159–169 (2008)

21. Satake, Y., Unno, H., Yanagi, H.: Probabilistic inference for predicate constraint satisfaction. In: The Thirty-Fourth AAAI Conference on Artificial Intelligence, AAAI 2020, The Thirty-Second Innovative Applications of Artificial Intelligence

Conference, IAAI 2020, The Tenth AAAI Symposium on Educational Advances in Artificial Intelligence, EAAI 2020, pp. 1644–1651. AAAI Press (2020)

22. Tsukada, T.: On computability of logical approaches to branching-time property verification of programs. In: Hermanns, H., Zhang, L., Kobayashi, N., Miller, D. (eds.) LICS 2020: 35th Annual ACM/IEEE Symposium on Logic in Computer Science, Saarbrücken, Germany, 8–11 July 2020, pp. 886–899. ACM (2020). https://doi.org/10.1145/3373718.3394766

23. Viswanathan, M., Viswanathan, R.: A higher order modal fixed point logic. In: Gardner, P., Yoshida, N. (eds.) CONCUR 2004. LNCS, vol. 3170, pp. 512–528. Springer, Heidelberg (2004). https://doi.org/10.1007/978-3-540-28644-8_33

24. Watanabe, K., Tsukada, T., Oshikawa, H., Kobayashi, N.: Reduction from branching-time property verification of higher-order programs to HFL validity checking. In: Hermenegildo, M.V., Igarashi, A. (eds.) Proceedings of the 2019 ACM SIGPLAN Workshop on Partial Evaluation and Program Manipulation, PEPM@POPL 2019, Cascais, Portugal, 14–15 January 2019, pp. 22–34. ACM (2019). https://doi.org/10.1145/3294032.3294077

Coefficient Synthesis for Threshold Automata

A. R. Balasubramanian[✉][iD]

Technische Universität München, Munich, Germany
bala.ayikudi@tum.de

Abstract. Threshold automata are a formalism for modeling fault-tolerant distributed algorithms. The main feature of threshold automata is the notion of a *threshold guard*, which allows us to compare the number of received messages with the total number of different types of processes. In this paper, we consider the coefficient synthesis problem for threshold automata, in which we are given a sketch of a threshold automaton (with the constants in the threshold guards left unspecified) and a specification and we want to synthesize a set of constants which when plugged into the sketch, gives a threshold automaton satisfying the specification. Our main result is that this problem is undecidable, even when the specification is a coverability specification and the underlying sketch is acyclic.

Keywords: Threshold automata · Coefficient synthesis · Presburger arithmetic with divisibility

1 Introduction

Threshold automata [7] are a formalism for modeling and analyzing parameterized fault-tolerant distributed algorithms. In this setup, an arbitrary but finite number of processes execute a given distributed protocol modeled as a threshold automaton. Verifying these systems amounts to proving that the given protocol is correct with respect to a given specification, irrespective of the number of agents executing the protocol. Many algorithms have been developed for verifying properties of threshold automata [2,4,7–10] and it is known that reachability for threshold automata is NP-complete [3].

In many formalisms for modeling distributed systems (like rendez-vous protocols [6] and reconfigurable broadcast networks [5]), the status of a transition being enabled or not depends only on a fixed number of processes, independent of the total number of participating processes. One of the central features that distinguishes threshold automata from such formalisms is the notion of a *threshold guard*. A threshold guard can be used to specify relationships between the

This project has received funding from the European Research Council (ERC) under the European Union's Horizon 2020 research and innovation programme under grant agreement No. 787367 (PaVeS).

number of messages received and the total number of participating processes, in order for a transition to be enabled. For example, if we let x be a variable counting the number of messages of a specified type, n be the number of participating processes and t be the maximum number of processes which can fail, then the guard $x \geq n/3 + t$ on a transition specifies that the number of messages received should be at least $n/3 + t$, in order for a process to execute this transition.

While the role of these guards is significant for the correctness of these protocols, they can also be unstable as small changes (and hence small calculation errors) in the coefficients of these guards can make a correct protocol faulty. (A concrete example of this phenomenon will be illustrated in the next section). For this reason, it would be desirable to automate the search for coefficients so that once the user gives a "sketch" of a threshold automaton (which only specifies the control flow but leaves out the arithmetic details) and a specification, we can compute a set of coefficient values, which when "plugged into" the sketch can satisfy the specification. With this motivation, the authors of [11] tackle this *coefficient synthesis problem* and provide theoretical and experimental results. They show that for a class of "sane" threshold automata, this problem is decidable and provide a CEGIS approach for synthesizing these coefficients. However, the decidability status of the coefficient synthesis problem for the general case has remained open so far.

In this paper, we prove that this problem is actually undecidable, hence settling the decidability status of this problem. We do this by giving a reduction from a sub-fragment of *Presburger arithmetic with divisibility*, for which the validity problem is known to be undecidable. Further, our result already shows that the coefficient synthesis problem is undecidable, even when the specification is a coverability specification of constant size and the underlying control-flow structure of the sketch automaton is acyclic.

Related Work. As mentioned before, the coefficient synthesis problem has already been studied in [11]. However, the decidability status of the general case was left open in that paper and here we show it is undecidable. A similar problem has also been studied for *parametric timed automata* [1], where the control flow of a timed automaton is given as input and we have to synthesize coefficients for the guards in order to satisfy a given reachability specification. The authors show that the problem is undecidable, already for timed automata with three clocks. They also show that it is decidable when the automaton has only one clock. Unlike clocks, the shared variables in our setting cannot be reset. Further, in our setting, variables can be compared with both the coefficients and other environment variables, which is not the case with parametric timed automata.

2   Preliminaries

Let $\mathbb{N}_{>0}$ be the set of positive integers and $\mathbb{N}$ be the set of non-negative integers.

```
1 var myval_i ∈ {0,1}
2 var accept_i ∈ {false, true} ← false
3
4 while true do (in one atomic step)
5  if myval_i = 1
6    and not sent ECHO before
7  then send ECHO to all
8
9  if received ECHO from at least
10    t + 1 distinct processes
11    and not sent ECHO before
12  then send ECHO to all
13
14  if received ECHO from at least
15    n − t distinct processes
16  then accept_i ← true
17 od
```

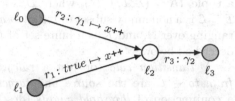

Fig. 1. Pseudocode of a reliable broadcast protocol from [13] for a correct process i, where n and t denote the number of processes, and an upper bound on the number of faulty processes. The protocol satisfies its specification (if $myval_i = 0$ for every correct process i, then no correct process sets its *accept* variable to *true*) if $t < n/3$.

Fig. 2. Threshold automaton from [9] modeling the body of the loop in the protocol from Fig. 1. Symbols γ_1, γ_2 stand for the threshold guards $x \geq (t + 1) - f$ and $x \geq (n - t) - f$, where n and t are as in Fig. 1, and f is the actual number of faulty processes. The shared variable x models the number of ECHO messages sent by correct processes. Processes with $myval_i = b$ (line 1) start in location ℓ_b (in green). Rules r_1 and r_2 model sending ECHO at lines 7 and 12. (Color figure online)

2.1 Threshold Automata

We introduce threshold automata, mostly following the definition and notations used in [2,3]. Along the way, we also illustrate the definitions on the example of Fig. 2 from [9], which is a model of the Byzantine agreement protocol of Fig. 1.

Environments. An *environment* is a tuple Env $= (\Pi, RC, N)$, where Π is a finite set of *environment variables* ranging over $\mathbb{N}$, $RC \subseteq \mathbb{N}^\Pi$ is a *resilience condition* over the environment variables, given as a linear formula, and $N: RC \to \mathbb{N}$ is a linear function called the *number function*. Intuitively, an assignment of Π determines the number of processes of different kinds (e.g. faulty) executing the protocol, and RC describes the admissible combinations of values of environment variables. Finally, N associates to a each admissible combination, the number of processes explicitly modeled. In a Byzantine setting, faulty processes behave arbitrarily, and so we do not model them explicitly; In the crash fault model, processes behave correctly until they crash and they must be modeled explicitly.

Example 1. In the threshold automaton of Fig. 2, the environment variables are n, f, and t, describing the number of processes, the number of (Byzantine) faulty processes, and the maximum possible number of faulty processes, respectively. The resilience condition is the constraint $n/3 > t \geq f$. The function N is given by $N(n, t, f) = n - f$, which is the number of correct processes.

Threshold Automata. A *threshold automaton* over an environment Env is a tuple $\mathsf{TA} = (\mathcal{L}, \mathcal{I}, \Gamma, \mathcal{R})$, where $\mathcal{L}$ is a finite set of *local states* (or *locations*), $\mathcal{I} \subseteq \mathcal{L}$ is a nonempty subset of *initial locations*, Γ is a finite set of *shared variables* ranging over $\mathbb{N}$, and $\mathcal{R}$ is a finite set of *transition rules* (or just *rules*), formally described below.

A *transition rule* (or just a *rule*) is a tuple $r = (from, to, \varphi, \vec{u})$, where $from, to \in \mathcal{L}$ are the *source* and *target* locations respectively, $\varphi \subseteq \mathbb{N}^{\Pi \cup \Gamma}$ is a conjunction of *threshold guards* (described below), and $\vec{u} \colon \Gamma \to \{0, 1\}$ is an *update*. We often let $r.from, r.to, r.\varphi, r.\vec{u}$ denote the components of r. Intuitively, r states that a process can move from *from* to *to* if the current values of Π and Γ satisfy φ, and when it moves, it updates the current valuation $\vec{g}$ of Γ by performing the update $\vec{g} := \vec{g} + \vec{u}$. Since all components of $\vec{u}$ are nonnegative, the values of shared variables never decrease. A *threshold guard* φ has one of the following forms: $b \cdot x \bowtie a_0 + a_1 \cdot p_1 + \ldots + a_{|\Pi|} \cdot p_{|\Pi|}$ where $\bowtie \in \{\geq, <\}$, $x \in \Gamma$ is a shared variable, $p_1, \ldots, p_{|\Pi|} \in \Pi$ are the environment variables, $b \in \mathbb{N}_{>0}$ and $a_0, a_1, \ldots, a_{|\Pi|} \in \mathbb{Z}$ are integer coefficients.

The underlying graph of a threshold automaton is the graph obtained by taking the vertices as the locations and connecting any two vertices with an edge as long as there is a rule between them. A threshold automaton is called *acyclic* if its underlying graph is acyclic.

Example 2. The threshold automaton from Fig. 2 is acyclic. The rule r_3 of this automaton has ℓ_2 and ℓ_3 as its source and target locations, $x \geq (n-t) - f$ as its guard, and does not increment any shared variable. On the other hand, the rule r_1 has ℓ_1 and ℓ_2 as its source and target locations, no guard (denoted by *true*) and increments the variable x.

Configurations and Transition Relation. A *configuration* of TA is a triple $\sigma = (\vec{\kappa}, \vec{g}, \mathbf{p})$ where $\vec{\kappa} \colon \mathcal{L} \to \mathbb{N}$ describes the number of processes at each location, and $\vec{g} \in \mathbb{N}^{\Gamma}$ and $\mathbf{p} \in RC$ are valuations of the shared variables and the environment variables respectively. In particular, $\sum_{\ell \in \mathcal{L}} \vec{\kappa}(\ell) = N(\mathbf{p})$ always holds. A configuration is *initial* if $\vec{\kappa}(\ell) = 0$ for every $\ell \notin \mathcal{I}$, and $\vec{g} = \vec{0}$. We often let $\sigma.\vec{\kappa}, \sigma.\vec{g}, \sigma.\mathbf{p}$ denote the components of σ.

A configuration $\sigma = (\vec{\kappa}, \vec{g}, \mathbf{p})$ *enables* a rule $r = (from, to, \varphi, \vec{u})$ if $\vec{\kappa}(from) > 0$, and $(\vec{g}, \mathbf{p})$ satisfies the guard φ, i.e., substituting $\vec{g}(x)$ for x and $\mathbf{p}(p_i)$ for p_i in φ yields a true expression, denoted by $\sigma \models \varphi$. If σ enables r, then there is a *step* from σ to the configuration $\sigma' = (\vec{\kappa}', \vec{g}', \mathbf{p}')$ given by, (i) $\mathbf{p}' = \mathbf{p}$, (ii) $\vec{g}' = \vec{g} + \vec{u}$, and (iii) $\vec{\kappa}' = \vec{\kappa} + \vec{v}_r$, where $\vec{v}_r = \vec{0}$ if *from* = *to* and otherwise, $\vec{v}_r(from) = -1$, $\vec{v}_r(to) = +1$, and $\vec{v}_r(\ell) = 0$ for all other locations ℓ. We let $\sigma \xrightarrow{r} \sigma'$ denote that

TA can move from σ to σ' using the rule r. We use $\sigma \to \sigma'$ to denote that $\sigma \xrightarrow{r} \sigma'$ for some rule r and we let $\sigma \xrightarrow{*} \sigma'$ denote the reflexive and transitive closure of the $\to$ relation. If $\sigma \xrightarrow{*} \sigma'$, then we say that there is a run from σ to σ' and also that σ' is reachable from σ.

Coverability. We will only be interested in *coverability specifications* throughout this paper. Later on, we will see how our result implies similar undecidability results for other specifications for threshold automata.

Let $\ell \in \mathcal{L}$ be a location. We say that a configuration σ covers ℓ if $\sigma(\ell) > 0$. We say that σ can cover a location ℓ if σ can reach a configuration σ' such that σ' covers ℓ. Finally, we say that TA can cover ℓ if some initial configuration of TA can cover ℓ. Hence, TA *cannot cover* ℓ if and only if every initial configuration of TA cannot cover ℓ. It is known that deciding whether a given threshold automaton can cover a given location is NP-complete [3].

Coefficient Synthesis. We now state the main problem that we are interested in this paper, namely the *coefficient synthesis* problem [11]. We begin by introducing the notion of *sketch threshold automata*.

Sketch Threshold Automata. An *indeterminate* is a variable that can take any integer value. In a typical threshold automaton, a guard is an inequality which can either be of the form $b \cdot x \geq a_0 + a_1 \cdot p_1 + \ldots a_{|\Pi|} \cdot p_{|\Pi|}$ or $b \cdot x < a_0 + a_1 \cdot p_1 + \ldots a_{|\Pi|} \cdot p_{|\Pi|}$ with $b \in \mathbb{N}_{>0}$ and $a_1, \ldots, a_{|\Pi|} \in \mathbb{Z}$. A sketch threshold automaton (or simply a sketch) is the same as a threshold automaton, except that some of the $b, a_0, a_1, \ldots, a_{|\Pi|}$ terms in any guard of the automaton are now allowed to be *indeterminates*. Intuitively, a sketch threshold automaton completely specifies the control flow of the protocol, but leaves out the precise arithmetic details of the threshold guards.

Given a sketch TA and an integer assignment μ to the indeterminates, we let TA$[\mu]$ denote the threshold automaton obtained by replacing the indeterminates with their corresponding values in μ. The *coefficient synthesis problem* for threshold automata is now the following:

> *Given:* An environment Env, a sketch TA and a location ℓ of TA
> *Decide:* Whether there is an assignment μ to the indeterminates such that TA$[\mu]$ cannot cover ℓ.

Example 3. We consider the threshold automaton from Fig. 2. As mentioned in the text under Fig. 1, if no (correct) process initially starts at ℓ_1, then no process can ever reach ℓ_3. This implies that if we remove the location ℓ_1 in the threshold automaton of Fig. 2, then the modified threshold automaton TA$'$ will never be able to cover ℓ_3.

We can now convert TA$'$ into a sketch, by replacing the guard γ_1 with $x \geq (t + a) - f$, where a is an indeterminate. When $a = 1$, we get TA$'$ and so it will never cover ℓ_3. However, when $a = 0$, the resulting automaton can cover

ℓ_3. Indeed if we set $n = 6, t = f = 1$ and if all the $N(6, 1, 1) = 6 - 1 = 5$ processes start at ℓ_0 initially, then the guard γ_1 will always be true and so all the 5 processes can move to ℓ_2, thereby setting the value of x to 5. At this point, the guard γ_2 becomes true and so all the processes can move to ℓ_3. This indicates that very small changes in the coefficients can make a protocol faulty.

Our main result is that

Theorem 1. *The coefficient synthesis problem is undecidable, even for acyclic threshold automata.*

3   Undecidability of Coefficient Synthesis

We now prove Theorem 1. To do so, we first consider a restricted version of this problem, called the *non-negative coefficient synthesis problem*, in which given a tuple (Env, TA, ℓ), we want to find a *non-negative assignment* μ to the indeterminates so that the resulting automaton TA$[\mu]$ does not cover ℓ. We first show that the non-negative coefficient synthesis problem is undecidable. Then, we reduce this problem to the coefficient synthesis problem, thereby achieving the desired result.

3.1   Presburger Arithmetic with Divisibility

We prove that the non-negative coefficient synthesis problem is undecidable by giving a reduction from the validity problem for a restricted fragment of Presburger arithmetic with divisibility, which is known to be undecidable. We describe this fragment by mostly following the definitions and notations given in [12].

Presburger arithmetic (PA) is the first-order theory over $\langle \mathbb{N}, 0, 1, +, < \rangle$ where $+$ and $<$ are the standard addition and order operations over the natural numbers $\mathbb{N}$ with constants 0 and 1 interpreted in the usual way. We can, in a straightforward manner, extend our syntax with the following abbreviations: $\leq, =, \geq, >$ and $ax = \sum_{1 \leq i \leq a} x$ where $a \in \mathbb{N}$ and x is a variable. A *linear polynomial* is an expression of the form $\sum_{1 \leq i \leq n} a_i x_i + b$ where each x_i is a variable and each a_i and b belong to $\mathbb{N}$. An *atomic formula* is a formula of the form $p(\mathbf{x}) \bowtie q(\mathbf{x})$ where p and q are linear polynomials over the variables $\mathbf{x}$ and $\bowtie \in \{<, \leq, =, >, \geq\}$.

Presburger arithmetic with divisibility (PAD) is the extension of PA obtained by adding a divisibility predicate $|$ which is interpreted as the usual divisibility relation among numbers. For the purposes of this paper, we restrict ourselves to the $\forall \exists_R \text{PAD}^+$ fragment of PAD, i.e., we shall only consider statements of the form

$$\forall x_1, \ldots, x_n \, \exists y_1, \ldots, y_m \bigvee_{i \in I} \left(\bigwedge_{(j,k) \in S_i} (x_j | y_k) \wedge \bigwedge_{l \in B_i} A_l(x_1, \ldots, x_n, y_1, \ldots, y_m) \right)$$

$$(1)$$

where I, S_i, B_i are finite sets of indices and each A_l is a quantifier-free atomic PA formula. It is known that checking if such a statement is true is undecidable [12]. We now prove the undecidability of the non-negative coefficient synthesis problem by a reduction from this problem.

Remark 1. PAD as defined here allows us to quantify the variables only over the natural numbers, whereas in [12] the undecidability result is stated for the variant where the variables are allowed to take integer values. However, the same proof given in [12] allows us to prove the undecidability result over the natural numbers as well.

Remark 2. In our definition of $\forall\exists_R\text{PAD}^+$, we only allow divisibility constraints of the form $x_j|y_k$. In [12], divisibility constraints of the form $f(\mathbf{x})|g(\mathbf{x}, \mathbf{y})$ were allowed, where f and g are any linear polynomials. This does not pose a problem, because of the fact that $\forall x_1, \ldots, x_n \, \exists y_1, \ldots, y_m \, f(\mathbf{x})|g(\mathbf{x}, \mathbf{y})$ is true if and only if $\forall x_1, \ldots, x_n, z \, \exists y_1, \ldots, y_m, z' \, (z \neq f(\mathbf{x})) \vee (z = f(\mathbf{x}) \wedge z' = g(\mathbf{x}, \mathbf{y}) \wedge z|z')$. Because of this identity, it is then clear that any formula in the $\forall\exists_R\text{PAD}^+$ fragment as defined in [12] can be converted into a formula in our fragment without changing its validity.

3.2   The Reduction

Let $\xi(x_1, \ldots, x_n, y_1, \ldots, y_m)$ be a formula of the form 1. The set of atomic formulas of ξ is the set comprising each quantifier-free atomic PA formula in ξ and all the divisibility constraints of the form $x_j|y_k$ that appear in ξ. The desired reduction now proceeds in two stages.

First Stage: The Environment. We begin by defining the environment $\text{Env} = (\Pi, RC, N)$. We will have m environment variables $t_1, \ldots, t_m$, with each t_i intuitively corresponding to the variable y_i in ξ. Further, for every atomic formula A of ξ which is a divisibility constraint, we will have an environment variable d_A. Finally, we will have an environment variable z, which will intuitively denote the total number of participating processes.

The resilience condition RC will be the trivial condition *true*. The linear function $N : RC \to \mathbb{N}$ is taken to be $N(\Pi) = z$. Hence, the total number of processes executing the threshold automaton will be z.

Second Stage: The Indeterminates and the Sketch. For each variable x_i of ξ, we will have an indeterminate s_i. Before we proceed with the description of the sketch, we make some small remarks.

Remark 3. Throughout the reduction, a *simple* configuration of a sketch will mean a configuration C such that 1) there is a **unique location** ℓ with $C(\ell) > 0$ and 2) $C(v) = 0$ for every shared variable v, i.e., all the processes of C are in exactly one location and the value of each shared variable is 0.

We now proceed with the description of the sketch. Throughout the reduction, we let **s** denote the set of indeterminates $s_1, \ldots, s_n$ and **t** denote the set of environment variables $t_1, \ldots, t_m$. The sketch will now be constructed in three phases, which are as follows.

First Phase. For each atomic formula A of ξ, we will construct a sketch TA_A. TA_A will have a single shared variable v_A. We now have two cases:

- Suppose A is of the form $x_j | y_k$ for some $j \in \{1, \ldots, n\}$ and $k \in \{1, \ldots, m\}$. Then, corresponding to A, we construct the sketch in Fig. 3.

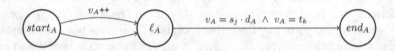

Fig. 3. Sketch for the first case

- Suppose A is of the form $f(\mathbf{x}, \mathbf{y}) \bowtie g(\mathbf{x}, \mathbf{y})$ where f and g are linear polynomials and $\bowtie \in \{<, \leq, =, >, \geq\}$. Then, corresponding to A, we construct the sketch in Fig. 4.

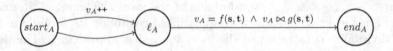

Fig. 4. Sketch for the second case

Remark 4. Notice that any assignment to the variables of **x** (resp. **y**) can be interpreted in a straightforward manner as an assignment to **s** (resp. **t**) and also vice versa. We will use this convention throughout the reduction.

Now let us give an intuitive idea behind the construction of these gadgets. Intuitively, in both these cases, all the processes initially start at $start_A$. Then each process either takes the top transition and increments v_A or takes the bottom transition and does not increment any variable. Ultimately, this would lead to a point where all the processes are now at ℓ_A. Then, in the first case, the guard from ℓ_A to end_A essentially checks that s_j divides t_k and in the second case, the guard from ℓ_A to end_A essentially checks that $f(\mathbf{s}, \mathbf{t}) \bowtie g(\mathbf{s}, \mathbf{t})$. By the previous remark, the variables **s** (resp. **t**) can be thought of as corresponding to the variables **x** (resp. **y**) and so this means that a process can reach end_A iff A can be satisfied. We now proceed to formalize this intuition.

Lemma 1. *Let X and Y be assignments to the variables **x** and **y** respectively. Then $A(X, Y)$ is true if and only if there is a simple configuration C of $\mathsf{TA}_A[X]$ with $C(start_A) > 0$ and $C(\mathbf{t}) = Y$ such that it can cover end_A.*

Proof. ($\Rightarrow$): Assume that $A(X,Y)$ is true.

- Suppose A is a divisibility constraint of the form $x_j | y_k$. Let q be such that $X(x_j) \cdot q = Y(y_k)$ and let C be the (unique) simple configuration given by $C(start_A) = C(z) = Y(y_k) + 1, C(t_k) = Y(y_k)$ and $C(d_A) = q$.
- Suppose A is of the form $f(\mathbf{x}, \mathbf{y}) \bowtie g(\mathbf{x}, \mathbf{y})$. Let C be the (unique) simple configuration given by $C(start_A) = C(z) = f(X,Y) + 1$ and $C(\mathbf{t}) = Y$.

The reason for having a "+1" in the definition of $C(start_A)$ is so that we are guaranteed to have at least one process to begin with.

From C, we proceed as follows: We move exactly one process from $start_A$ to ℓ_A by using the rule which increments nothing and we move all the other processes, one by one, from $start_A$ to ℓ_A by using the rule which increments v_A. This leads to a configuration C' such that $C'(v_A) = C'(z) - 1 = C(z) - 1$. Because we assume that $A(X,Y)$ is true, it follows that at C', the outgoing rule from ℓ_A is enabled. Hence, we can now move a process from ℓ_A into end_A, thereby covering end_A.

($\Leftarrow$): Assume that C is a simple configuration of $\mathsf{TA}_A[X]$ with $C(start_A) > 0$ and $C(\mathbf{t}) = Y$ such that from C it is possible to cover end_A. Let ρ be a run from C which covers end_A. By construction of TA_A, it must mean that the outgoing rule from ℓ_A is fired at some point along the run and so its guard must be enabled at some configuration C' along the run. Note that $C'(\mathbf{t}) = C(\mathbf{t})$, since the environment variables never change their value along a run.

Now, suppose A is of the form $x_j | y_k$. This means that we have $X(s_j) \cdot C'(d_A) = C'(t_k)$. Since $X(s_j) = X(x_j), C'(t_k) = C(t_k) = Y(t_k)$, this implies that $X(x_j)$ divides $Y(t_k)$ and so $A(X,Y)$ is true. On the other hand, suppose A is of the form $f(\mathbf{x}, \mathbf{y}) \bowtie g(\mathbf{x}, \mathbf{y})$. Since $C'(\mathbf{t}) = C(\mathbf{t}) = Y$, this implies that $f(X,Y) \bowtie g(X,Y)$ and so $A(X,Y)$ is true. $\square$

Second Phase. Let $\{\xi_i\}_{i \in I}$ be the set of subformulas of ξ such that $\xi = \forall x_1, \ldots, x_n \exists y_1, \ldots, y_m \bigvee_{i \in I} \xi_i$, i.e., the subformula ξ_i is the disjunct corresponding to the index i in the formula ξ. Let $A_i^1, \ldots, A_i^{l_i}$ be the set of atomic formulas appearing in ξ_i. We construct a sketch threshold automaton TA_{ξ_i} in the following manner: We take the sketches $\mathsf{TA}_{A_i^1}, \ldots, \mathsf{TA}_{A_i^{l_i}}$ from the first phase and then for every $1 \le j \le l_i - 1$, we add a rule which connects $end_{A_i^j}$ to $start_{A_i^{j+1}}$, which neither increments any shared variable nor has any threshold guards. This is illustrated in Fig. 5 for the case of $l_i = 3$.

To prove a connection between the constructed gadget and the formula ξ_i, we first need to state a property of the gadget. We begin with a definition.

Definition 1. *Let C, C' be two configurations of $\mathsf{TA}_{\xi_i}[X]$ for some assignment X and let $A \in \{A_i^1, \ldots, A_i^{l_i}\}$. We say that $C \preceq_A C'$ if $v \in \{v_A, d_A, \mathbf{t}\}$ implies that $C(v) = C'(v)$ and $v \in \{start_A, \ell_A, end_A, z\}$ implies that $C(v) \le C'(v)$.*

By construction of TA_{ξ_i}, the following *monotonicity property* is clear.

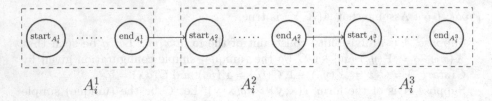

Fig. 5. Example sketch for the second phase

Proposition 1 (Monotonicity). *Let X be an assignment to the indeterminates and let $C \xrightarrow{r} C'$ be a step in $\mathsf{TA}_{\xi_i}[X]$ such that the rule r belongs to $\mathsf{TA}_{A_i^j}$ for some j. Then for every D such that $C \preceq_{A_i^j} D$, there exists a D' such that $D \xrightarrow{r} D'$ is a step in $\mathsf{TA}_{\xi_i}[X]$ and $C' \preceq_{A_i^j} D'$.*

We now have the following proof which asserts the correctness of our construction.

Lemma 2. *Let X and Y be assignments to the variables $\mathbf{x}$ and $\mathbf{y}$ respectively. Then $\xi_i(X,Y)$ is true if and only if there is a simple configuration C of $\mathsf{TA}_{\xi_i}[X]$ with $C(start_{A_i^1}) > 0$ and $C(\mathbf{t}) = Y$ such that it can cover $end_{A_i^{l_i}}$.*

Proof. ($\Rightarrow$): Suppose $\xi_i(X,Y)$ is true. This means that $A_i^j(X,Y)$ is true for every $1 \le j \le l_i$. By Lemma 1, for every $1 \le j \le l_i$, there exists a simple configuration C_j of $\mathsf{TA}_{A_i^j}[X]$ with $C_j(start_{A_i^j}) > 0$ and $C_j(\mathbf{t}) = Y$ such that C_j can cover $end_{A_i^j}$. For each j, let $C_j \xrightarrow{*} C_j'' \xrightarrow{r_j} C_j'$ be a shortest run from C_j which covers $end_{A_i^j}$. By definition $C_j''(end_{A_i^j}) = 0$ and $C_j'(end_{A_i^j}) > 0$. This means that the (unique) outgoing rule from $\ell_{A_i^j}$ is enabled at C_j'' and r_j is in fact, this rule. This also implies that the only difference between C_j'' and C_j' is that a process has moved from $\ell_{A_i^j}$ to $end_{A_i^j}$. In particular, the shared variables and the environment variables do not change their values during this step and so the guards along the rule r_j are true at C_j' as well.

Let $Z = \max\{C_j(z) : 1 \le j \le l_i\}$. Further, for each A_i^j which is a divisibility constraint of the form $x_k \mid y_l$, let $W_{A_i^j} = Y(y_l)/X(x_k)$. Let D_1 be the configuration given by $D_1(\mathbf{t}) = Y, D_1(d_A) = W_A$ for every $A \in \{A_i^1, \ldots, A_i^{l_i}\}$ which is a divisibility constraint, $D_1(z) = D_1(start_{A_i^1}) = Z$ and $D_1(v) = 0$ for every other v. Note that $C_1 \preceq_{A_i^1} D_1$.

We will now show the following by induction: For any $1 \le j \le l_i$, there is a configuration D_j which is reachable from D_1 such that $C_j \preceq_{A_i^j} D_j, D_j(start_{A_i^j}) = Z$ and $D_j(v_{A_i^k}) = 0$ for every $k \ge j$. The base case of $j = 1$ is trivial. Assume that we have already shown it for some j and we now want to prove it for $j + 1$. By existence of the run $C_j \xrightarrow{*} C_j'$ and because of the monotonicity property, there is a run $D_j \xrightarrow{*} D_j'$ such that $C_j' \preceq_{A_i^j} D_j'$. Since the guards of the outgoing rule from $\ell_{A_i^j}$ are enabled at C_j', it follows that it is also enabled at D'_j. We now do the following: From D'_j, we first move all the processes at $start_{A^j_i}$ to $\ell_{A^j_i}$ by means of the rule which increments nothing. From there we move all the processes at $\ell_{A^j_i}$ to $end_{A^j_i}$ and then to $start_{A^{j+1}_i}$. This results in a configuration D_{j+1} which satisfies the claim.

By induction, this means that we can reach D_{l_i} from D_1. By the monotonicity property, we can cover $end_{A^{l_i}_i}$ from D_{l_i}.

($\Leftarrow$): Suppose there is a simple configuration C of $\mathsf{TA}_{\xi_i}[X]$ with $C(start_{A^1_i}) > 0$ and $C(\mathbf{t}) = Y$ such that it can cover $end_{A^{l_i}_i}$. Let $C \xrightarrow{*} C'$ be such a run. By construction of TA_{ξ_i}, this implies that there must be configurations $C_1, \ldots, C_{l_i}$ along this run such that at each C_j, the outgoing rule from $\ell_{A^j_i}$ must be enabled. Hence, this means that if A^j_i is a formula of the form $x_k \mid y_l$, then $X(s_k) \cdot C_j(d_{A^j_i}) = C_j(t_l)$ and if A^j_i is a formula of the form $f_j(\mathbf{x}, \mathbf{y}) \bowtie g_j(\mathbf{x}, \mathbf{y})$, then $f_j(X(\mathbf{s}), C_j(\mathbf{t})) \bowtie g_j(X(\mathbf{s}), C_j(\mathbf{t}))$. Since environment variables do not change their values along a run, this implies that in the former case, $X(x_k)|Y(y_l)$ and in the latter case, $f_j(X, Y) \bowtie g_j(X, Y)$. Hence, $A^j_i(X, Y)$ is true for every j and so ξ_i is true. □

Third Phase. The final sketch threshold automaton TA is constructed as follows: TA will have a copy of each of the TA_{ξ_i} and in addition it will also have two new locations $start$ and end. Then, for each index $i \in I$, TA will have two rules, one of which goes from $start$ to $start_{A^1_i}$ and the other from $end_{A^{l_i}_i}$ to end. Both of these rules do not increment any variable and do not have any guards. Intuitively, these two rules correspond to choosing the disjunct ξ_i from the formula ξ. This is illustrated in Fig. 6 for the case when the index set $I = \{i, j, k\}$.

Setting the initial set of locations of TA to be $\{start\}$, we have the following.

Proposition 2. *Let X and Y be assignments to the variables $\mathbf{x}$ and $\mathbf{y}$ respectively. Then $\xi(X, Y)$ is true iff some initial configuration C with $C(\mathbf{t}) = Y$ can cover the location end in $\mathsf{TA}[X]$.*

Proof. ($\Rightarrow$): Suppose $\xi(X, Y)$ is true. Then $\xi_i(X, Y)$ is true for some i. By Lemma 2, there exists a simple configuration C of $\mathsf{TA}_{\xi_i}[X]$ such that $C(start_{A^1_i}) > 0$ and $C(\mathbf{t}) = Y$ which can cover $end_{A^{l_i}_i}$. Consider the initial configuration D in TA which is the same as C except that $D(start_{A^1_i}) = 0$ and $D(start) = C(start_{A^1_i})$. By construction of $\mathsf{TA}[X]$, we can make D reach C. Since we can cover $end_{A^{l_i}_i}$ from C in $\mathsf{TA}_{\xi_i}[X]$, we can also cover it in $\mathsf{TA}[X]$. Once we can cover $end_{A^{l_i}_i}$, we can also cover end.

($\Leftarrow$): Suppose there is some initial configuration C with $C(\mathbf{t}) = Y$ which can cover the location end in $\mathsf{TA}[X]$. Let $C \xrightarrow{*} C'$ be such a run covering end. By construction, there must be an index $i \in I$ and configurations $C_1, C_2, \ldots, C_{l_i}$ along this run such that at each C_j, the outgoing rule from $\ell_{A^j_i}$ is enabled. Similar to the argument from Lemma 2, we can show that each $A^j_i(X, Y)$ is true and so $\xi_i(X, Y)$ is true, which implies that $\xi(X, Y)$ is true. □

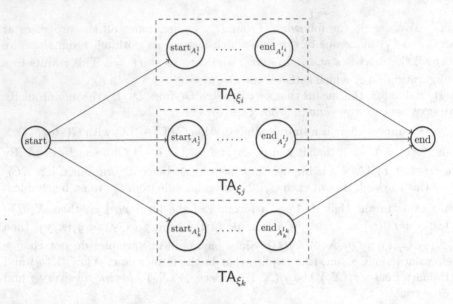

Fig. 6. Example sketch for the third phase

It then follows that $\forall \mathbf{x}\ \exists \mathbf{y}\ \xi(\mathbf{x}, \mathbf{y})$ is true iff for every assignment X of the indeterminates of TA, there exists an initial configuration C such that C can cover *end* in TA$[X]$. Since TA is acyclic, it follows that

Theorem 2. *The non-negative coefficient synthesis problem for threshold automata is undecidable, even for acyclic threshold automata.*

Example 4. We illustrate the above reduction on an example. Suppose we have the formula

$$\forall x_1, x_2, \exists y_1, y_2\ (x_1|y_1) \vee (x_2|y_1 \wedge x_1 = 2x_2 + y_2) \tag{2}$$

Let A, B, and C denote the subformulas $x_1|y_1$, $x_2|y_1$, and $x_1 = 2x_2 + y_2$ respectively. For this formula, our reduction produces the sketch given in Fig. 7.

Here s_1, s_2 are indeterminates corresponding to x_1, x_2 and t_1, t_2 are environment variables corresponding to y_1, y_2. Notice that the formula is true, because if x_1 is assigned the value a and x_2 is assigned the value b, then we can always set y_1 to a and y_2 to b, which will always make the first disjunct true. Similarly, in the sketch threshold automaton, if μ is any assignment to the indeterminates, then by letting C be the (unique) initial configuration such that $C(t_1) = \mu(s_1)$, $C(t_2) = \mu(s_2)$, $C(z) = \mu(s_1) + 1$ and $C(d_A) = C(d_B) = 1$, we can cover *end*$_A$ from C and so we can also cover *end* from C.

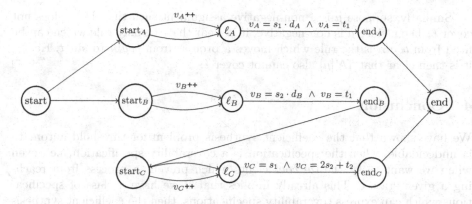

Fig. 7. Sketch for formula 2

3.3 Wrapping Up

We can now reduce the non-negative coefficient synthesis problem to the coefficient synthesis problem.

Theorem 3. *The coefficient synthesis problem for threshold automata is undecidable, even for acyclic threshold automata.*

Proof. Let $(\mathsf{Env}, \mathsf{TA}, \ell)$ be an instance of the non-negative coefficient synthesis problem. Without loss of generality, we can assume that TA is acyclic and has only a single initial location *start*. This is because we have shown earlier that the non-negative coefficient synthesis problem is already undecidable for inputs satisfying this property.

Let X be the set of indeterminates of TA. We now add a new location *begin* and a new shared variable *check*. *check* will have the invariant that it will never be incremented by any of the rules. Now, from *begin* we add $|X| + 1$ rules as follows: First, we add a rule from *begin* to *start* which neither increments any variable nor has any guards. Then for each indeterminate $x \in X$, we add a rule from *begin* to ℓ which has the guard *check* $> x$. Notice that since *check* is never incremented, it will always have the value 0 and so the guard *check* $> x$ will be true if and only if x takes a negative value. Finally, we set the new initial location to be *begin* and we let this new sketch threshold automaton be TA'. Notice that TA' is acyclic.

We will now prove that $(\mathsf{Env}, \mathsf{TA}, \ell)$ is a yes instance of the non-negative coefficient synthesis problem if and only if $(\mathsf{Env}, \mathsf{TA}', \ell)$ is a yes instance of the coefficient synthesis problem.

Notice that if μ is an assignment to X such that $\mu(x) < 0$ for some $x \in X$, then it is possible to move a process from *begin* to ℓ. Hence, if μ assigns a negative value to some indeterminate, then there is at least one run from an initial configuration in $\mathsf{TA}'[\mu]$ which covers ℓ. Hence, if no initial configuration of $\mathsf{TA}'[\mu]$ can cover ℓ, then μ has to be a non-negative assignment. But then it is easy to see that no initial configuration of $\mathsf{TA}[\mu]$ can cover ℓ as well.

Similarly, suppose μ is a non-negative assignment such that TA$[\mu]$ does not cover ℓ. Then, since μ is non-negative, it is clear that the only rule which can be fired from *begin* is the rule which moves a process from *begin* to *start*. Hence, it is then clear that TA$'[\mu]$ also cannot cover ℓ.   □

4   Conclusion

We have shown that the coefficient synthesis problem for threshold automata is undecidable, when the specification is a coverability specification, i.e., even when we want to synthesize coefficients which prevent a process from reaching a given state ℓ. This already implies that if we have a class of specifications which can express coverability specifications, then the coefficient synthesis problem for threshold automata generalized to that class is also undecidable. For instance, this implies that coefficient synthesis for specifications from the ELTL$_{FT}$ logic [9], which has been used to express various properties of threshold automata obtained from distributed algorithms, is also undecidable. Further our reduction implies that the coefficient synthesis problem is undecidable even for threshold automata with an acyclic control-flow structure.

References

1. Alur, R., Henzinger, T.A., Vardi, M.Y.: Parametric real-time reasoning. In: Kosaraju, S.R., Johnson, D.S., Aggarwal, A. (eds.) Proceedings of the Twenty-Fifth Annual ACM Symposium on Theory of Computing, San Diego, CA, USA, 16–18 May 1993, pp. 592–601. ACM (1993). https://doi.org/10.1145/167088.167242
2. Balasubramanian, A.R.: Parameterized complexity of safety of threshold automata. In: Saxena, N., Simon, S. (eds.) FSTTCS 2020. LIPIcs, vol. 182, pp. 37:1–37:15 (2020). https://doi.org/10.4230/LIPIcs.FSTTCS.2020.37
3. Balasubramanian, A.R., Esparza, J., Lazić, M.: Complexity of verification and synthesis of threshold automata. In: Hung, D.V., Sokolsky, O. (eds.) ATVA 2020. LNCS, vol. 12302, pp. 144–160. Springer, Cham (2020). https://doi.org/10.1007/978-3-030-59152-6_8
4. Bertrand, N., Konnov, I., Lazić, M., Widder, J.: Verification of randomized consensus algorithms under round-rigid adversaries. In: CONCUR. LIPIcs, vol. 140, pp. 33:1–33:15 (2019)
5. Delzanno, G., Sangnier, A., Zavattaro, G.: Parameterized verification of ad hoc networks. In: Gastin, P., Laroussinie, F. (eds.) CONCUR 2010. LNCS, vol. 6269, pp. 313–327. Springer, Heidelberg (2010). https://doi.org/10.1007/978-3-642-15375-4_22
6. German, S.M., Sistla, A.P.: Reasoning about systems with many processes. J. ACM **39**(3), 675–735 (1992)
7. Konnov, I., Veith, H., Widder, J.: On the completeness of bounded model checking for threshold-based distributed algorithms: reachability. Inf. Comput. **252**, 95–109 (2017)
8. Konnov, I., Widder, J.: ByMC: Byzantine model checker. In: Margaria, T., Steffen, B. (eds.) ISoLA 2018. LNCS, vol. 11246, pp. 327–342. Springer, Cham (2018). https://doi.org/10.1007/978-3-030-03424-5_22

9. Konnov, I.V., Lazic, M., Veith, H., Widder, J.: A short counterexample property for safety and liveness verification of fault-tolerant distributed algorithms. In: POPL 2017, pp. 719–734 (2017)
10. Kukovec, J., Konnov, I., Widder, J.: Reachability in parameterized systems: all flavors of threshold automata. In: CONCUR, pp. 19:1–19:17 (2018)
11. Lazić, M., Konnov, I., Widder, J., Bloem, R.: Synthesis of distributed algorithms with parameterized threshold guards. In: OPODIS. LIPIcs, vol. 95, pp. 32:1–32:20 (2017)
12. Pérez, G.A., Raha, R.: Revisiting parameter synthesis for one-counter automata. In: CSL 2022. LIPIcs (2022). https://doi.org/10.4230/LIPIcs.CSL.2022.33
13. Srikanth, T., Toueg, S.: Simulating authenticated broadcasts to derive simple fault-tolerant algorithms. Distrib. Comput. **2**, 80–94 (1987). https://doi.org/10.1007/BF01667080

Subsequences in Bounded Ranges:
Matching and Analysis Problems

Maria Kosche[(✉)] [iD], Tore Koß[iD], Florin Manea[iD], and Viktoriya Pak

Göttingen University, Göttingen, Germany
{maria.kosche,tore.koss,florin.manea,viktoriya.pak}@cs.uni-goettingen.de

Abstract. In this paper, we consider a variant of the classical algorithmic problem of checking whether a given word v is a subsequence
of another word w. More precisely, we consider the problem of deciding, given a number p (defining a range-bound) and two words v and
w, whether there exists a factor $w[i : i + p - 1]$ (or, in other words,
a range of length p) of w having v as subsequence (i.e., v occurs as
a subsequence in the bounded range $w[i : i + p - 1]$). We give matching upper and lower quadratic bounds for the time complexity of this
problem. Further, we consider a series of algorithmic problems in this
setting, in which, for given integers k, p and a word w, we analyse the
set p-$\mathrm{Subseq}_k(w)$ of all words of length k which occur as subsequence of
some factor of length p of w. Among these, we consider the k-universality
problem, the k-equivalence problem, as well as problems related to absent
subsequences. Surprisingly, unlike the case of the classical model of subsequences in words where such problems have efficient solutions in general, we show that most of these problems become intractable in the new
setting when subsequences in bounded ranges are considered. Finally,
we provide an example of how some of our results can be applied to
subsequence matching problems for circular words.

Keywords: Subsequences · Bounded range · Matching problems ·
Analysis problems · Algorithms · Fine grained complexity

1 Introduction

A word u is a subsequence of a string w if there exist (possibly empty) strings
$v_1, \ldots, v_{\ell+1}$ and $u_1, \ldots, u_\ell$ such that $u = u_1 \ldots u_\ell$ and $w = v_1 u_1 \ldots v_\ell u_\ell v_{\ell+1}$. In
other words, u can be obtained from w by removing some of its letters. In this
paper, we focus on words occurring as subsequences in bounded ranges of a word.

The notion of subsequences appears in various areas of computer science. For
instance, in automata theory, the theory of formal languages, and logics, it is
used in connection with piecewise testable languages [45–47,66,67], or subword
order and downward closures [7,39,51,52,71]. Naturally, subsequences appear in

The work of Maria Kosche and Tore Koß was supported by the DFG-grant MA 5725/2–
1. The work of Florin Manea was supported by the DFG-grant MA 5725/3–1.

the area of combinatorics and algorithms on words [5,26,28,53,54,58,62–64], but they are also used for modelling concurrency [14,61,65], as well as in database theory (especially in connection with *event stream processing* [3,38,72,73]). Nevertheless, a series of classical, well-studied, and well-motivated combinatorial and algorithmic problems deal with subsequences. Some are stringology problems, such as the longest common subsequence problem [1,2,4,12,13] or the shortest common supersequence problem [56]), but there are also problems related to the study of patterns in permutations, such as increasing subsequences or generalizations of this concept [8,9,17,20,23,27,37,42].

In general, one can split the algorithmic questions related to the study of subsequences in two large classes. The first class of problems is related to *matching* (or searching), where one is interested in deciding whether a given word u occurs as a subsequence in another (longer) word w. The second class contains *analysis problems*, which are focused on the investigation of the sets $\mathrm{Subseq}_k(w)$ of all subsequences of length k of a given string w (of course, we can also remove the length restriction, and investigate the class of all subsequences of w). In this setting, one is interested, among other problems, in deciding the *k-universality problem*, i. e., whether $\mathrm{Subseq}_k(w) = \Sigma^k$, where w and k are given, or the *equivalence problem*, i. e., whether $\mathrm{Subseq}_k(w) = \mathrm{Subseq}_k(u)$, where w, u and k are given. In the general case of subsequences, introduced above, the matching problem can be solved trivially by a greedy algorithm. The case of analysis problems is more interesting, but still well-understood (see, e. g., [59,60]). For instance, the equivalence problem, which was introduced by Imre Simon in his PhD thesis [66], was intensely studied in the combinatorial pattern matching community (see [16,25,35,40,68,69] and the references therein). This problem was optimally solved in 2021 [36]. The universality problem was also intensely studied (see [5,18] and the references therein); to this end, we will also recall the work on absent subsequences in words [50], where the focus is on minimal strings (w. r. t. length or the subsequence relation) which are not contained in $\mathrm{Subseq}(u)$.

Getting closer to the topic of this paper, let us recall the following two scenarios related to the motivation for the study of subsequences, potentially relevant in the context of reachability and avoidability problems. Assume that w is some string (or stream) we observe, which may represent, on the one hand, the trace of some computation or, on the other hand, and in a totally different framework, the DNA-sequence describing some gene. Deciding whether a word v is a subsequence of w can be interpreted as deciding, in the first case, whether the events described by the trace v occurred during the longer computation described by w in the same order as in v, or, in the second case, if there is an alignment between the sequence of nucleotides v and the longer sequence w. However, in both scenarios described above, it seems unrealistic to consider occurrences of v in w where the positions of w matching the first and last symbol of v, respectively, are very far away from each other. It seems indeed questionable, for instance, whether considering an alignment of DNA-sequences v and w where the nucleotides of v are spread over a factor of w which is several times longer than v itself is still meaningful. Similarly, when observing a computation, one might be more interested in its recent history (and the sequences of events occurring there), rather than analysing the entire computation. Moreover, the fact that in many practical scenarios (including those mentioned above) one has to process streams, which, at any moment, can only be partly accessed by our algorithms, enforces even more the idea that the case where one is interested in subsequences occurring arbitrarily in a given string is less realistic and less useful than the case where one is interested in the subsequences occurring in bounded ranges of the respective string (which can be entirely accessed and processed at any moment by our algorithms).

Hence, we consider in this paper the notion of p-subsequence of a word. More precisely, a word v is a p-subsequence of w if there exists a factor (or bounded range) $w[i : i+p-1]$ of w, of length p, such that v is a subsequence of $w[i : i+p-1]$. In this framework, we investigate both matching and analysis problems.

Our Results. With respect to the matching problem, we show that checking, for a given integer p, whether a word v is a p-subsequence of another word w can be done in $\mathcal{O}(|w||v|)$ time and $\mathcal{O}(|v|)$ space, and show that this is optimal conditional to the Orthogonal Vectors Hypothesis. With respect to the analysis problem, we show that the problem of checking, for given integers k and p and word w, whether there exists a word v of length k which is not a p-subsequence of w is NP-hard. Similarly, checking, for given integer p and words v and w, whether the sets of p-subsequences of w and v are not equal is also NP-hard. These results are complemented by conditional lower bounds for the time complexity of algorithms solving them. Several results related to the computation of absent p-subsequences are also shown. Interestingly, we show that checking if a word is a shortest absent p-subsequence of another word is NP-hard, while checking if a word is a minimal absent p-subsequence of another word can be done in quadratic time (and this is optimal conditional to the Orthogonal Vectors Hypothesis). We end the paper with a series of results related to subsequences in circular words. Among other results, we also close an open problem from [41] that is related to the computation of minimal representatives of circular words.

Related Works. Clearly, considering properties of bounded ranges (or factors of bounded length) in words can be easily related to the study of sliding window algorithms [30–34] or with algorithms in the streaming model [6,21]. For our algorithmic results, we discuss their relation to results obtained in those frameworks. Moreover, the notion of subsequences with gap constraints was recently introduced and investigated [19]. In that case, one restricts the occurrences of a subsequence v in a word w, by placing length (and regular) constraints on the factors of w which are allowed to occur between consecutive symbols of the subsequence when matched in w. Our framework (subsequences in bounded range) can be, thus, seen as having a general length constraint on the distance between the position of w matching the first symbol of v and the position of w matching the last symbol of v, in the occurrences of v inside w.

Structure of the Paper. We first give a series of preliminary definitions. Then we discuss the matching problem. Further, we discuss the analysis problems and the problems related to absent subsequences. We then discuss the case of subsequences of circular words. We end with a conclusions section. For space reasons, missing proofs and additional comments are only presented in the full version of this paper [49].

2 Basic Definitions

Let $\mathbb{N}$ be the set of natural numbers, including 0. For $m, n \in \mathbb{N}$, we define the range (or interval) of natural numbers lower bounded by m and upper bounded by n as $[m : n] = \{m, m + 1, \ldots, n\}$. An alphabet Σ is a non-empty finite set of symbols (called letters). A *string (or word)* is a finite sequence of letters from Σ, thus an element of the free monoid Σ^*. Let $\Sigma^+ = \Sigma^* \setminus \{\varepsilon\}$, where ε is the empty string. The *length* of a string $w \in \Sigma^*$ is denoted by $|w|$. The ith letter of $w \in \Sigma^*$ is denoted by $w[i]$, for $i \in [1 : |w|]$. Let $|w|_a = |\{i \in [1 : |w|] \mid w[i] = a\}|$; let $\text{alph}(w) = \{x \in \Sigma \mid |w|_x > 0\}$ be the smallest subset $S \subseteq \Sigma$ such that $w \in S^*$. For $m, n \in \mathbb{N}$, with $m \leq n$, we define the range (or factor) of w between positions m and n as $w[m : n] = w[m]w[m + 1] \ldots w[n]$. If $m > n$, then $w[m : n]$ is the empty word. Also, by convention, if $m < 1$, then $w[m : n] = w[1 : n]$, and if $n > |w|$, then $w[m : n] = w[m : |w|]$. A factor $u = w[m : n]$ of w is called a *prefix* (respectively, *suffix*) of w if $m = 1$ (respectively, $n = |w|$).

The powers of a word w are defined as: $w^0 = \varepsilon$ and $w^{k+1} = ww^k$, for $k \geq 0$. We define w^ω as the right infinite word which has w^n as prefix for all $n \geq 0$. The positive integer $p \leq |w|$ is a period of a word w if w is a prefix of $w[1 : p]^\omega$. Let $w = w_1 \ldots w_n$ (for some $w_1, \ldots, w_n \in \Sigma$) and $p \in \mathbb{N}$. Then $w^{\frac{p}{n}} = w^{\lfloor \frac{p}{n} \rfloor} w'$, where $w' = w_1 \ldots w(p \bmod n)$.

We recall the notion of subsequence.

Definition 1. *A word v is a subsequence of length k of w (denoted $v \leq w$), where $|w| = n$, if there exist positions $1 \leq i_1 < i_2 < \ldots < i_k \leq n$, such that $v = w[i_1]w[i_2] \cdots w[i_k]$. The set of all subsequences of w is denoted by $\text{Subseq}(w)$.*

When v is not a subsequence of w we also say that v is *absent* from w.

The main concept discussed here is that of p-subsequence, introduced next.

Definition 2. *1. A word v is a p-subsequence of w (denoted $v \leq_p w$) if there exists $i \leq |w| - p + 1$ such that v is a subsequence of $w[i : i + p - 1]$.*
2. For $p \in \mathbb{N}$ and $w \in \Sigma^*$, we denote the set of all p-subsequences of w by p-$\text{Subseq}(w) = \{v \in \Sigma^* \mid v \leq_p w\}$. Furthermore, for $k \in \mathbb{N}$, we denote the set of all p-subsequences of length k of w by p-$\text{Subseq}_k(w)$.

Extending the notions of absent subsequences introduced in [50], we now define the notion of (shortest and minimal) absent p-subsequences in a word.

Definition 3. *The word v is an absent p-subsequence of w if $v \notin p$-$\text{Subseq}(w)$. We also say v is p-absent from w. The word v is a p-SAS (shortest absent p-subsequence) of w if v is an absent p-subsequence of w of minimal length. The word v is a p-MAS (minimal absent p-subsequence) of w if v is an absent p-subsequence of w but all subsequences of v are p-subsequences of w.*

Note that, in general, any shortest absent $|w|$-subsequence of a word w (or, simply, shortest absent subsequence of w, denoted SAS) has length $\iota(w) + 1$, where $\iota(w) = \max\{k \mid \mathrm{alph}(w)^k \subseteq \mathrm{Subseq}(w)\}$ is the universality index of w [5].

Computational Model. In general, the problems we discuss here are of algorithmic nature. The computational model we use to describe our algorithms is the standard unit-cost RAM with logarithmic word size: for an input of size N, each memory word can hold $\log N$ bits. Arithmetic and bitwise operations with numbers in $[1 : N]$ are, thus, assumed to take $\mathcal{O}(1)$ time. In all the problems, we assume that we are given a word w or two words w and u, with $|w| = n$ and $|v| = m$ (so the size of the input is $N = n + m$), over an alphabet $\Sigma = \{1, 2, \ldots, \sigma\}$, with $2 \le |\Sigma| = \sigma \le n + m$. That is, we assume that the processed words are sequences of integers (called letters or symbols), each fitting in $\mathcal{O}(1)$ memory words. This is a common assumption in string algorithms: the input alphabet is said to be *an integer alphabet*. For more details see, e. g., [15].

Our algorithmic results (upper bounds) are complemented by a series of lower bounds. In those cases, we show that our results hold already for the case of constant alphabets. That is, they hold already when the input of the problem is restricted to words over an alphabet $\Sigma = \{1, 2, \ldots, \sigma\}$, with $\sigma \in \mathcal{O}(1)$.

Complexity Hypotheses. As mentioned, we are going to show a series of conditional lower bounds for the time complexity of the considered problems. Thus, we now recall some standard computational problems and complexity hypotheses regarding them, respectively, on which we base our proofs of lower bounds.

The *Satisfiability problem for formulas in conjunctive normal form*, in short CNF-SAT, gets as input a Boolean formula F in conjunctive normal form as a set of clauses $F = \{c_1, c_2, \ldots, c_m\}$ over a set of variables $V = \{v_1, v_2, \ldots, v_n\}$, i. e., for every $i \in [m]$, we have $c_i \subseteq \{v_1, \neg v_1, \ldots, v_n, \neg v_n\}$. The question is whether F is satisfiable. By k-CNF-SAT, we denote the variant where $|c_i| \le k$ for all $i \in [m]$.

The *Orthogonal Vectors problem* (OV for short) gets as input two sets A, B each containing n Boolean-vectors of dimension d, where $d \in \omega(\log n)$. The question is whether there exist two vectors $\vec{a} \in A$ and $\vec{b} \in B$ which are orthogonal, i. e., $\vec{a}[i] \cdot \vec{b}[i] = 0$ for every $i \in [d]$.

We shall use the following algorithmic hypotheses based on CNF-SAT and OV that are common for obtaining conditional lower bounds in fine-grained complexity. In the following, poly is any fixed polynomial function:

- *Exponential Time Hypothesis* (ETH) [44,55]: 3-CNF-SAT cannot be solved in time $2^{o(n)} \mathrm{poly}(n + m)$.
- *Strong Exponential Time Hypothesis* (SETH) [43,70]: For every $\epsilon > 0$ there exists k such that k-CNF-SAT cannot be decided in $\mathcal{O}(2^{n(1-\epsilon)} \mathrm{poly}(n))$ time.

The following result, which essentially formulates the Orthogonal Vectors Hypothesis (OVH), can be shown (see [10,11,70]).

Lemma 1. *OV cannot be solved in $\mathcal{O}(n^{2-\epsilon}\mathrm{poly}(d))$ time for any $\epsilon > 0$, unless SETH fails.*

3   Matching Problems

We first consider the following problem.

Subsequence Matching in Bounded Range, pSubSeqMatch
Input: Two words u and w over Σ and an integer p, with $|u| = m$ and
$|w| = n$, and $m \leq p \leq n$.
Question: Decide whether $u \leq_p w$.

Theorem 1. pSubSeqMatch *can be solved in* $\mathcal{O}(mn)$ *time.*

Proof. We present an algorithm which solves pSubSeqMatch in $\mathcal{O}(mn)$ time
and works in a streaming fashion w. r. t. the word w. More precisely, our
algorithm scans the letters of w one by one, left to right (i. e., in the order
$w[1], w[2], \ldots, w[n]$), and after scanning the letter $w[t]$, for $t \geq p$, it decides
whether u is a subsequence of the bounded range $w[t - p + 1 : t]$ (i. e., our
algorithm works as a sliding window algorithm, with fixed window size, and the
result for the currently considered window is always calculated before the next
letter is read).

Let us explain how this algorithm works. We maintain an array $A[\cdot]$ with
m elements such that the following invariant holds. For $t \geq 0$: after the tth
letter of w is scanned and A is updated, $A[i]$ is the length of the shortest suffix
of $w[t - p + 1 : t]$ which contains $u[1 : i]$ as a subsequence (or $A[i] = +\infty$ if
$w[t - p + 1 : t]$ does not contain $u[1 : i]$ as a subsequence). Note that, before
reading the letter $w[t]$, for all $t \in [1 : n]$, the array A, if correctly computed, has
the property that $A[i] \leq A[i + 1]$, for all $i \in [1 : m]$.

Initially, we set $A[i] = +\infty$ for all $i \in [1 : m]$.

Let us see how the elements of A are updated when $w[t]$ is read. We first
compute $\ell \leftarrow |u|_{w[t]}$ and the positions $j_1, \ldots, j_\ell \in [1 : m]$ such that $u[j_h] = w[t]$.
Then, we compute an auxiliary array $B[\cdot]$ with m elements, in which we set
$B[j_h] \leftarrow A[j_h - 1] + 1$, for $h \in [1 : \ell]$. For $i \in [1 : m] \setminus \{j_1, \ldots, j_\ell\}$ we set
$B[i] \leftarrow A[i] + 1$. Intuitively, $B[i]$ is the length of the shortest suffix of $w[t - p : t]$
which contains $u[1 : i]$ as a subsequence. Then, we update $A[i]$, for $i \in [1 : m]$,
by setting $A[i] \leftarrow B[i]$ if $B[i] \leq p$ and $A[i] \leftarrow +\infty$, otherwise.

After performing the update of the array $A[\cdot]$ corresponding to the scanning
of letter $w[t]$, we decide that u occurs in the window $w[t - p + 1 : t]$ if (and only
if) $A[m] \leq p$.

Let us show that the algorithm is correct. The stated invariant clearly holds
before scanning the letters of w (i. e., after 0 letters were scanned). Assume that
the invariant holds after $f - 1$ letters of w were scanned. Now, we will show that
it holds after f letters were scanned. Assume that, for some prefix $u[1 : i - 1]$
of u, we have that $w[f' : f - 1]$ is the shortest suffix of $w[f - p : f - 1]$ which
contains $u[1 : i - 1]$ as subsequence. Next, we scan letter $w[f]$. If $w[f] = u[i]$,
then $w[f' : f]$ is the shortest suffix of $w[f - p : f]$ which contains $u[1 : i]$ as
subsequence (otherwise, there would exist a suffix of $w[f - p : f - 1]$ that is
shorter than $w[f' : f - 1]$ which contains $u[1 : i - 1]$ as subsequence). So, it is

correct to set $B[i] \leftarrow A[i-1]+1$, when $w[f] = u[i]$. Otherwise, if $w[f] \neq u[i]$, we note that the shortest suffix of $w[f-p:f]$ which contains $u[1:i]$ as subsequence starts on the same position as the shortest suffix of $w[f-p:f-1]$ which contains $u[1:i]$ as subsequence. Therefore, $B[i] \leftarrow A[i]+1$ is also correct (as we compute the length of these shortest suffixes w. r. t. the currently scanned position of w). Then, we simply update A to only keep track of those suffixes of the currently considered range of size p, i. e., $w[f-p+1:f]$.

The algorithm runs in $\mathcal{O}(nm)$ time. Indeed, for each scanned letter $w[f]$ we perform $\mathcal{O}(m)$ operations. Moreover, the space complexity of the algorithm is $\mathcal{O}(m)$, as we only maintain the arrays A and B. So, the statement holds.   □

As stated in the proof of Theorem 1, the algorithm we presented can be seen as an algorithm in the sliding window model with window of fixed size p (see [30,31,34]). More precisely, we scan the stream w left to right and, when the tth letter of the stream is scanned, we report whether the window $w[t-p+1:p]$ contains u as a subsequence. In other words, we report whether the string $w[t-p+1:p]$ is in the regular language $L_u = \{v \mid u \leq v\}$. The problem of checking whether the factors of a stream scanned by a sliding window are in a regular language was heavily investigated, see [29] and the references therein. In particular, from the results of [31] it follows that, for a constant u (i. e., u is not part of the input), the problem of checking whether the factors of a stream scanned by a sliding window are in the language L_u cannot be solved using $o(\log p)$ bits when the window size is not changing and equals p. We note that our algorithm is optimal from this point of view: if u is constant and, thus, $m \in \mathcal{O}(1)$, our algorithm uses $\mathcal{O}(\log p)$ bits to store the arrays A and B.

We can show that our algorithm is also optimal (conditional to OVH) also from the point of view of time complexity.

Theorem 2. pSubSeqMatch *cannot be solved in time* $\mathcal{O}(n^h m^g)$, *where* $h+g = 2-\epsilon$ *with* $\epsilon > 0$, *unless OVH fails.*

Proof. Let (A,B) be an instance of OV with $A = \{a_1,\dots,a_n\} \subset \{0,1\}^d$ and $B = \{b_1,\dots,b_n\} \subset \{0,1\}^d$. Furthermore let $v = (v[1],v[2],\dots,v[d])$ for every $v \in A \cup B$. We represent a_i and b_j by the strings

$$\psi_A(a_i) = \psi_A(a_i[1])\psi_A(a_i[2])\cdots\psi_A(a_i[d]),$$
$$\psi_B(b_j) = \psi_B(b_j[1])\psi_B(b_j[2])\cdots\psi_B(b_j[d])$$

where $\psi_A(x) = \begin{cases} 01\# & \text{if } x = 0, \\ 00\# & \text{if } x = 1 \end{cases}$ and $\psi_B(y) = y\#$ for $y \in \{0,1\}$.

Claim I. a_i and b_j are orthogonal if and only if $\psi_B(b_j)$ occurs in $\psi_A(a_i)$ as a subsequence.

Proof (of Claim I*).* Since $|\psi_A(a_i)|_\# = |\psi_B(b_j)|_\# = d$ holds, $\psi_B(b_j)$ is a subsequence of $\psi_A(a_i)$ if and only if $\psi_B(b_j[k])$ occurs in $\psi_A(a_i[k])$ for all $k \in [d]$. We note that, for $x,y \in \{0,1\}$, $\psi_B(y)$ is absent from $\psi_A(x)$ if and only if $x = y = 1$.

Hence, $\psi_B(b_j)$ is a subsequence of $\psi_A(a_i)$ if and only if $a_i[k] \cdot b_j[k] = 0$ for all $k \in [d]$. That is, $\psi_B(b_j)$ is a subsequence of $\psi_A(a_i)$ if and only if a_i is orthogonal to b_j.   (End of the proof of Claim I)   □

With ψ_A and ψ_B we construct two words $W, U \in \{0, 1, \#, [,]\}^*$ representing A and B as follows:

$$W = [\psi_A(\mathbb{1})][\psi_A(\mathbb{0})][\psi_A(a_1)][\psi_A(\mathbb{0})][\psi_A(a_2)][\psi_A(\mathbb{0})] \dots [\psi_A(a_n)][\psi_A(\mathbb{0})][\psi_A(\mathbb{1})]$$
$$U = [\psi_B(\mathbb{1})][\psi_B(b_1)][\psi_B(b_2)] \dots [\psi_B(b_n)][\psi_B(\mathbb{1})]$$

where $\mathbb{0}$ (respectively, $\mathbb{1}$) stands for the all-zero (respectively, all-one) vector of size d. We will occasionally omit ψ_A and ψ_B and call $[\psi_A(v)]$ and $[\psi_B(v)]$ $[v]$-blocks or, more generally, $[\cdot]$-blocks if it is clear whether it occurs in W or in U. As such, $[\psi_A(\mathbb{0})]$ is called a zero-block in the following, while a $[v]$-block is called a non-zero-block if and only if $v \neq \mathbb{0}$. Thus, the encodings of U and W include two $[\mathbb{1}]$-blocks at the beginning and at the end of U and W, respectively, and a zero-block after each non-zero-block of W (excluding the $[\mathbb{1}]$-block at the end).

Remark 1. Since U starts with $[$, if U is a subsequence of any bounded range of length $|W|$ of W^2, then U is a subsequence of a bounded range of length $|W|$ of W^2 starting with $[$.

Next we show that the instance (A, B) of OV is accepted if and only if the instance $u = U$, $w = W^2$ and $p = |W|$ of pSubSeqMatch is accepted.

Claim II. There are orthogonal vectors $a_i \in A$ and $b_j \in B$ if and only if U is a $|W|$-subsequence of W^2.

Proof (of Claim II). Firstly, suppose a_i and b_j are orthogonal. Then $[\psi_B(b_j)]$ occurs in $[\psi_A(a_i)]$. Since $|W| = (2n + 3) \cdot |[\psi_A(v)]|$ for any $v \in \{0, 1\}^d$, we can choose a bounded range (until the end of this proof the reader may safely assume every bounded range to be a bounded range of length $|W|$ of W^2) containing $2n + 2$ $[\cdot]$-blocks around $[\psi_A(a_i)]$. Furthermore, every bounded range starting with $[$ contains exactly $n + 1$ zero-blocks. Hence, we choose a bounded range containing j zero-blocks to the left of $[\psi_A(a_i)]$ and $n - j + 1$ zero-blocks to the right of $[\psi_A(a_i)]$. If $j \leq i$, we match the $[b_j]$-block in U against the first occurrence of the $[a_i]$-block in W^2 and choose the bounded range starting at the first occurrence of $[\psi_A(a_{i-j})]$. If $j > i$, we match the $[b_j]$-block in U against the second occurrence of the $[a_i]$-block in W^2 and choose the bounded range starting at the first occurrence of $[\psi_A(a_{n+i-j+1})]$. In both cases there is one zero-block to the left (respectively, right) of $[\psi_A(a_i)]$ for one $[\mathbb{1}]$-block and each $[b_k]$-block for $1 \leq k < j$ (respectively, $j < k \leq n$). Hence, U is a $|W|$-subsequence of W^2.

For the inverse implication, suppose that a_i is not orthogonal to b_j for all $1 \leq i, j \leq n$. By Claim I, no $[b_j]$-block occurs in any $[a_i]$-block as a subsequence, hence the $[\cdot]$-blocks of U only occur in zero-blocks of W^2. By Remark 1, it suffices to show that U does not occur in a bounded range starting with $[$. Each of those has $n + 1$ zero-blocks but U has $n + 2$ non-zero-blocks. Thus, U is $|W|$-absent from W^2.   (End of the proof of Claim II)   □

Finally, we note that $|W| = (2n + 3)(3d + 2) \in \mathcal{O}(n \cdot poly(d))$ and $|U| = (n + 2)(2d + 2) \in \mathcal{O}(n \cdot poly(d))$ and so an algorithm deciding pSubSeqMatch in time $\mathcal{O}(|W|^h|U|^g)$, with $h + g = 2 - \epsilon$, could also be used to decide OV in time $\mathcal{O}(n^{2-\epsilon}poly(d))$, which is not possible by OVH. Hence, Theorem 2 holds.   □

4   Analysis Problems

This section covers several decision problems regarding the set p-Subseq_k defined for an input word. More precisely, we approach problems related to the universality of this set, the equivalence of the respective sets for two words, as well as problems related to minimal and shortest absent subsequences w. r. t. these sets. For these problems we give respective hardness results and fine-grained conditional lower bounds. We start with the following problem.

---

k-Non-Universality in Bounded Range, kpNonUniv

Input: A word w over Σ and integers k, p, with $|w| = n$, and $k \leq p \leq n$.

Question: Decide whether p-$\mathrm{Subseq}_k(w) \neq \Sigma^k$.

Let us observe that if $|w| < k\sigma$, where $\sigma = |\Sigma|$, then we can trivially conclude by the definition of the universality of a word that $\mathrm{Subseq}_k(w) \neq \Sigma^k$, so p-$\mathrm{Subseq}_k(w) \neq \Sigma^k$ as well. Therefore, to avoid the trivial inputs of kpNonUniv we will assume that $|w| \geq k\sigma$.

To show that kpNonUniv is NP-complete, we first examine a related problem given in [57]. We need some preliminaries. A partial word over an alphabet Σ is a string from $(\Sigma \cup \{\Diamond\})^*$. In such a partial word u, we have defined positions (those positions i for which $u[i] \in \Sigma$) and undefined positions (those positions i for which $u[i] = \Diamond$); intuitively, while the letters on the defined positions are fixed, the $\Diamond$ can be replaced by any letter of the alphabet Σ and, as such, a partial word actually describes a set of (full) words over Σ^*, all of the same length as u. This is formalized as follows. If u and v are partial words of equal length, then u is contained in v, denoted by $u \subseteq v$, if $u[i] = v[i]$, for all defined positions i (i. e., all positions i such that $u[i] \in \Sigma$). Moreover, the partial words u and v are compatible, denoted by $u \uparrow v$, if there exists a full word w such that $u \subseteq w$ and $v \subseteq w$. In this framework, we can define the problem partialWordsNonUniv, which we will use in our reductions.

Partial words non-universality, partialWordsNonUniv

Input: A list of partial words $S = \{w_1, \ldots, w_k\}$ over $\{0, 1\}$, each partial word having the same length L

Question: Decide whether there exists a word $v \in \{0, 1\}^L$ such that v is not compatible with any of the partial words in S.

The first part of following result was shown in [57] via a reduction from 3-CNF-SAT, and it can be complemented by a conditional lower bound.

Theorem 3. partialWordsNonUniv *is NP-complete and cannot be solved in subexponential time* $2^{o(L)} \, poly(L, n)$ *unless* ETH *fails.*

Based on the hardness of `partialWordsNonUniv`, we continue by showing that the k-Non-Universality in Bounded Range Problem is also NP-hard.

Theorem 4. `kpNonUniv` *is NP-hard and cannot be solved in subexponential time* $2^{o(k)}\operatorname{poly}(k,n)$ *unless ETH fails.*

Proof (Sketch). Let $S = \{w_1, \ldots, w_k\}$ be a set of partial words, each of length L, defining an instance of `partialWordsNonUniv`. We reduce this to an instance of `kpNonUniv` as follows. For every partial word $w_i \in S$, construct $u_i = u_i^1 u_i^2 \cdots u_i^L$ where

$$u_i^j = \begin{cases} 0\# & \text{if } w_i[j] = 0 \\ 1\# & \text{if } w_i[j] = 1 \qquad\qquad \text{for } j \in [L], \\ 01\# & \text{if } w_i[j] = \Diamond \end{cases}$$

Now, we define:

$$V = \#^{2L}(001101\#^{2L})^{L-1},$$
$$U = \#^{4L^2} u_1 \#^{4L^2} u_2 \#^{4L^2} \cdots \#^{4L^2} u_k \#^{4L^2},$$

and set $W = VU$.

We can show (see the paper's full version [49]) that the instance of `kpNonUniv` defined by the input word W and $k = 2L$ and $p = |V|$ is accepted if and only if the instance of `partialWordsNonUniv` defined by the set S is accepted.

So, we have a valid reduction from `partialWordsNonUniv` to `kpNonUniv`. Moreover, as this reduction can be performed in polynomial time and we have $k = 2L$, we also obtain an ETH lower bound for `kpNonUniv`. That is, `kpNonUniv` cannot be solved in subexponential time $2^{o(k)}\operatorname{poly}(k,n)$ unless ETH fails.   □

This hardness result is complemented by the following algorithmic result.

Remark 2. Note that `kpNonUniv` can be trivially solved in $\mathcal{O}(\sigma^k \operatorname{poly}(k,n))$ by a brute-force algorithm that simply checks for all words from Σ^k whether they are in $p\text{-Subseq}_k(w)$. For $\sigma \in \mathcal{O}(1)$, this algorithm runs in $2^{o(k)}\operatorname{poly}(k,n)$.

Now, when looking at a related analysis problem, we consider two different words w and v, and we want to check whether both words are equivalent w.r.t. their respective sets of p-subsequences of length k.

---

k-Non-Equivalence w.r.t. Bounded Ranges, kpNonEquiv
Input: Two words w and v over Σ and integers k, p, with $|w| = n$, $|v| = m$, and $k \le p \le m, n$.
Question: Decide whether $p\text{-Subseq}_k(w) \neq p\text{-Subseq}_k(v)$.

We can now state the following theorem.

Theorem 5. `kpNonEquiv` *is NP-hard and cannot be solved in subexponential time* $2^{o(k)}\operatorname{poly}(k,n,m)$ *unless ETH fails.*

Again, a matching upper bound is immediate.

Remark 3. kpNonEquiv can be trivially solved in $\mathcal{O}(\sigma^k \operatorname{poly}(k,n))$ by a brute-force algorithm that looks for a word from Σ^k which separates $p\text{-Subseq}_k(w)$ and $p\text{-Subseq}_k(v)$. For $\sigma \in \mathcal{O}(1)$, this algorithm runs in $2^{o(k)} \operatorname{poly}(k,n)$.

A natural problem arising in the study of the sets $p\text{-Subseq}_k(w)$, for $k \leq p \leq |w|$, is understanding better which are the strings missing from this set. To that end, we have introduced in Sect. 2 the notions of shortest and minimal absent p-subsequences, p-SAS and p-MAS, respectively.

We first focus on shortest absent subsequences in bounded ranges.

Non-Shortest Absent Subsequence w. r. t. Bounded Ranges, pNonPSAS
Input: Two words w and v over Σ and integer p, with $|w| = n$, $|v| = m$, and $m \leq p \leq n$.
Question: Decide whether v is not a p-SAS of w..

We can show the following result.

Theorem 6. pNonPSAS *is NP-hard and cannot be solved in subexponential time* $2^{o(k)} \operatorname{poly}(k,n,m)$ *unless* ETH *fails.*

Now let us examine minimal absent subsequences in bounded ranges, p-MAS. First we give an algorithm to check whether a string is a p-MAS of another string.

Minimal Absent Subsequences w. r. t. Bounded Ranges, pMAS
Input: Two words w and v over Σ and integer p, with $|w| = n$, $|v| = m$, and $m \leq p \leq n$.
Question: Decide whether v is a p-MAS of w.

In this case, we obtain a polynomial time algorithm solving this problem.

Theorem 7. pMAS *can be solved in time* $\mathcal{O}(nm)$, *where* $|v| = m$, $|w| = n$.

Similarly to the proof of Theorem 1, we propose an algorithm (see paper's full version [49]) which can be seen as working in the sliding window model, with window of fixed size p. If, as in the case of the discussion following Theorem 1, we assume u (and m) to be constant, we obtain a linear time algorithm. However, its space complexity, measured in memory words, is $\mathcal{O}(p)$ (as we need to keep track, in this case, of entire content of the window). In fact, when m is constant, it is easy to obtain a linear time algorithm using $\mathcal{O}(1)$ memory words (more precisely, $\mathcal{O}(\log p)$ bits of space) for this problem: simply try to match u and all its subsequences of length $(m-1)$ in w simultaneously, using the algorithm from Theorem 1. Clearly, u is a p-MAS if and only if u is not a subsequence of w, but its subsequence of length $m-1$ are. However, the constant hidden by the $\mathcal{O}$-notation in the complexity of this algorithm is proportional with m^2. It remains open whether there exists a (sliding window) algorithm for pMAS both running in $\mathcal{O}(mn)$ time (which we will show to be optimal, conditional to OVH) and using only $\mathcal{O}(\log p)$ bits (which is also optimal for sliding window algorithms, see appendix of the full version of this paper [49]).

Complementing the discussion above, we can show that it is possible to construct in linear time (see paper's full version [49]), for words u, w and integer $p \in \mathbb{N}$, a string w' such that deciding whether u is a p-MAS of w' is equivalent to deciding whether u is a p-subsequence of w, so solving pSubSeqMatch for the input words u and w. Hence, the lower bound from Theorem 2 carries over, and the algorithm in Theorem 7 is optimal (conditional to OVH) from the time complexity point of view.

Theorem 8. pMAS *cannot be solved in time* $\mathcal{O}(n^h m^g)$ *where* $h + g = 2 - \epsilon$ *with* $\epsilon > 0$, *unless OVH fails.*

5   Application: Subsequence Matching in Circular Words

An interesting application of the framework developed in the previous setting is related to the notion of circular words.

We start with a series of preliminary definitions and results.

Intuitively, a circular word w_o is a word obtained from a (linear) word $w \in \Sigma^*$ by linking its first symbol after its last one as shown in Fig. 1.

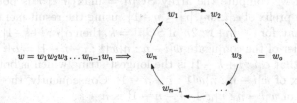

Fig. 1. The circular word w_o, defined via the (linear) word w.

Formally, two words $u, w \in \Sigma^*$ are conjugates (denoted $u \sim w$) if there exist $x, y \in \Sigma^*$ such that $u = xy$ and $w = yx$. The conjugacy relation $\sim$ is an equivalence relation on Σ^*, and the circular word w_o is defined as the equivalency class of w with respect to $\sim$ (i. e., the set of all words equivalent to w w. r. t. $\sim$). Clearly, for a word w of length n, the equivalence class of w with respect to $\sim$ has at most n elements. It is worth investigating how we can represent circular words. To this end, we use the following definition from [41].

Definition 4. *A pair* $(u, n) \in \Sigma^* \times \mathbb{N}$ *is a representation of the circular word* w_o *if* $|u| \leq n, n = |w|$ *and* $u^{\frac{n}{|u|}} \in w_o$. *A minimal representation of a circular word* w_o *is a pair* (u, n) *such that* (u, n) *is a representation of* w_o *and for each other pair* (u', n) *which represents* w_o *we have that* $|u| < |u'|$ *or* $|u| = |u'|$ *and* u *is lexicographically smaller than* u'.

As an example, $(baa, 5)$ is a representation of the circular word $baaba_o$, because $(baa)^{5/3} = baaba$, but its minimal representation is $(ab, 5)$. Indeed,

$ab^{5/2} = ababa$, which is a conjugate of $baaba$, as $baaba = ba \cdot aba$ and $ababa = aba \cdot ba$. As another example, both $(ababa, 12)$ and $(babaa, 12)$ are representations of the circular word $aababaababaa_\circ$, but $(ababa, 12)$ is the minimal representation.

Clearly, every circular word $w_\circ$ has a minimal representation, and an open problem from [41] is how efficiently can the minimal representation of a circular word $w_\circ$ be computed. We solve this open problem (only proved in the full version of this paper, as it is not directly connected to the main topic of this paper [49]).

Theorem 9. *Given a word w of length n, the minimal representative (v, n) of $w_\circ$ and a conjugate $w[i : n]w[1 : i-1] = v^{n/|v|}$ of w can be computed in $\mathcal{O}(n)$ time.*

Proof. We begin by referring to [22]. In Lemma 5 of the respective paper it is shown that, for a word u of length n, we can compute in $\mathcal{O}(n)$ time the values $SC[i] = \max\{|r| \mid r$ is both a suffix of $w[1..i-1]$ and a prefix of $w[i..n]\}$. The proof of that lemma can be directly adapted to prove the following result: given a word u of length n and an integer $\Delta \leq n$, we can compute in $\mathcal{O}(n)$ time the values $SC[i] = \max\{|r| \mid r$ is both a suffix of $w[1..i-1]$ and a prefix of $w[i..n], |r| \leq \Delta\}$. We will use this in the following.

Let us now prove the statement of our theorem. We consider the word $\alpha = www$, and we compute the array $SC[i] = \max\{|r| \mid r$ is both a suffix of $\alpha[1..i-1]$ and a prefix of $\alpha[i..3n], |r| \leq n-1\}$, using the result mentioned above. Now, we note that, for $i \in [n+1 : 2n]$ if $SC[i] = k$, then $\alpha[i : i+k-1]$ is the longest non-trivial border of the conjugate $w[i-n : n]w[1 : i-n-1] = \alpha[i : i+n-1]$ of w; that means that $\alpha[i : i+k-1]$ is the longest string which is both non-trivial suffix and prefix of $w[i-n : n]w[1 : i-n-1]$. Consequently, the length of the shortest period of $w[i-n : n]w[1 : i-n-1]$ is $n-k$.

In conclusion, we have computed for each conjugate $w[j : n]w[1 : j-1]$ of w its shortest period $n - SC[j+n]$. Further, we can sort these conjugates w.r.t. their shortest period using counting sort. In this way, we obtain a conjugate $w[j : n]w[1 : j-1]$ of w which has the shortest period among all the conjugates of w. In the case of multiple such conjugates $w[j_g : n]w[1 : j_g-1]$, with $g \in [1 : \ell]$ for some ℓ, we proceed as follows. We construct (in linear time) the suffix array of α (as in [48]), and set $j \leftarrow j_h$ where $j_h + n$ occurs as the first in the suffix array of α among all positions $j_g + n$, for $g \in [1 : \ell]$. Therefore, we obtain a conjugate $w' = w[j : n]w[1 : j-1]$ of w which has the shortest period among all the conjugates of w, and is lexicographically smaller than all other conjugates of w which have the same period. Moreover, for $p = n - SC[j+n]$, we have that $w' = (w'[1 : p])^{n/p}$.

The algorithm computing w' and its period runs, clearly, in $\mathcal{O}(n)$ time, as all its steps can be implemented in linear time. Our statement is, thus, correct. $\square$

This concludes the preliminaries part of this section.

In this framework, we define subsequences of circular words.

Definition 5. *A word v is a subsequence of a circular word $w_\circ$ ($v \leq_\circ w$) if and only if there exists a conjugate $w' = w[i+1..n]w[1..i]$ of w such that v is a subsequence of w'.*

This definition follows from [5] where one defines k-circular universal words as words $u \in \Sigma^*$ which have at least one conjugate u' whose set of subsequences of length k is Σ^k. In this setting, we can define the following problem.

Circular Subsequence Matching, $\texttt{SubSeqMatch}_\circ$
Input: Two words u and w over Σ, with $|u| = m$, $|w| = n$, and $m \leq n$.
Question: Decide whether $v \leq_\circ w$.

As the conjugates of a word w, of length n, are the factors of length n of ww, we immediately obtain the following result from Theorem 1.

Theorem 10. $\texttt{SubSeqMatch}_\circ$ can be solved in $\mathcal{O}(mn)$ time.

And furthermore, the proof of Theorem 2 shows that the following statement also holds.

Theorem 11. $\texttt{SubSeqMatch}_\circ$ cannot be solved in time $\mathcal{O}(n^h m^g)$, where $h + g = 2 - \epsilon$ with $\epsilon > 0$, unless OVH fails.

To conclude this paper, we consider an extension of the $\texttt{SubSeqMatch}_\circ$ which seems natural to us. We begin by noting that reading (or, more precisely, traversing all the positions of) a circular word $w_\circ$ can be interpreted as reading (respectively, going through) the letters written around a circle, as drawn, for instance, in Fig. 1. So, we can start reading the word at some point on this circle, then go once around the circle, until we are back at the starting point. Then, as in a loop, we could repeat reading (traversing) the word. So, it seems natural to ask how many times do we need to read/traverse a circular word $w_\circ$ until we have that a given word u is a subsequence of the word we have read/traversed.

Clearly, this problem is not well defined, as it depends on the starting point from which we start reading the circular word $w_\circ$. Let us consider an example. Consider the word $w = ababcc$. Now, if we consider the circular word $w_\circ$, and we start reading/traversing it from position one of w (i.e., we start reading $ababcc \cdot ababcc \cdot ababcc \cdot \ldots$) then we need to read/traverse twice the circular word to have that ca is a subsequence of the traversed word. But if we start reading/traversing the circular word on any position $i \geq 2$ of w (e.g., we start on position 3 and read $abccab \cdot abccab \cdot abccab \cdot \ldots$), then it is enough to traverse the circular word once to have that ca is a subsequence of the traversed word.

In this setting, there are two natural ways to fix this issue.

The first one is to define a canonical point of start for the traversal. A natural choice for this starting point is to consider a special position of the word such as, for instance, the positions where a minimal representative of $w_\circ$ occurs. To this end, a conjugate $u = w[i : n]w[1 : i - 1]$ of a word w of length n is called minimal rotation of w if $u = v^{n/|v|}$ and (v, n) is a minimal representative of $w_\circ$. We obtain the following problem (presented here as a decision problem).

Iterated Circular Subsequence Matching, $\texttt{ItSubSeqMatch}_\circ$
Input: An integer ℓ, a word v, and a word w, which defines the circular word $w_\circ$, over Σ, with $|v| = m$ and $|w| = n$, and $m \leq n$.
Question: Decide whether $v \leq u^\ell$, where u is a minimal rotation of w.

This problem is clearly well defined now, as if u and u' are minimal rotations of w, then $u = u'$. Moreover, this problem can be also formulated as a minimisation problem by simply asking for the smallest ℓ for which $\texttt{ItSubSeqMatch}_{\circ}$ with input (ℓ, v, w) can be answered positively.

The second way to solve the issue identified above is as follows.

Best Iterated Circular Subsequence Matching, $\texttt{BestItSubSeqMatch}_{\circ}$

Input: An integer ℓ, a word v, and a word w, which defines the circular word $w_{\circ}$, over Σ, with $|v| = m$ and $|w| = n$, and $m \leq n$.

Question: Decide whether there is a conjugate u of w such that $v \leq u^{\ell}$.

Clearly, this problem can be also formulated as a minimisation problem by simply asking for the smallest ℓ for which $\texttt{BestItSubSeqMatch}_{\circ}$ with input (ℓ, v, w) can be answered positively.

Our results regarding these two problems are summarized below.

Theorem 12. *1.* $\texttt{ItSubSeqMatch}_{\circ}$ *(and the related minimisation problem) can be solved in* $\mathcal{O}(\min(n\sigma + m, n + m \log \log n))$ *time, where* $\sigma = |\Sigma|$.

2. $\texttt{BestItSubSeqMatch}_{\circ}$ *(and the corresponding minimisation problem) can be solved in* $\mathcal{O}(nm)$ *time.*

3. $\texttt{BestItSubSeqMatch}_{\circ}$ *cannot be solved in time* $\mathcal{O}(n^h m^g)$, *where* $h + g = 2 - \epsilon$ *with* $\epsilon > 0$, *unless OVH fails.*

As a comment on the previous result, it remains open whether there are solutions for $\texttt{ItSubSeqMatch}_{\circ}$ which are more efficient than our algorithms, whose complexity is given in statement 1 of Theorem 12.

6   Conclusions

In this paper we have considered a series of classical matching and analysis problems related to the occurrences of subsequences in words, and extended them to the case of subsequences occurring in bounded ranges in words. In general, we have shown that the matching problem, where we simply check if a word is a subsequence of another word, becomes computationally harder in this extended setting: it now requires rectangular time. Similarly, problems like checking whether two words have the same set of subsequences of given length or checking whether a word contains all possible words of given length as subsequences become much harder (i.e., NP-hard as opposed to solvable in linear time) when we consider subsequences in bounded ranges instead of arbitrary subsequences. We have also analysed a series of problems related to absent subsequences in bounded ranges of words and, again, we have seen that this case is fundamentally different than the case of arbitrary subsequences. In general, our results paint a comprehensive picture of the complexity of matching and analysis problems for subsequences in bounded ranges.

As an application of our results, we have considered the matching problem for subsequences in circular words, where we simply check if a word u is a subsequence of any conjugate of another word w (i.e., is v a subsequence of the circular word w_o), and we have shown that this problem requires quadratic time. A series of other results regarding the occurrences of subsequences in circular words were discussed, but there are also a few interesting questions which remained open in this setting: What is the complexity of deciding whether two circular words have the same set of subsequences of given length? What is the complexity of checking whether a circular word contains all possible words of given length as subsequences? Note that the techniques we have used to show hardness in the the case of analysis problems for subsequences in bounded ranges of words do not seem to work in the case of circular words, so new approaches would be needed in this case.

References

1. Abboud, A., Backurs, A., Williams, V.V.: Tight hardness results for LCS and other sequence similarity measures. In: IEEE 56th Annual Symposium on Foundations of Computer Science, FOCS 2015, Berkeley, CA, USA, 17–20 Oct 2015, pp. 59–78 (2015). https://doi.org/10.1109/FOCS.2015.14
2. Abboud, A., Williams, V.V., Weimann, O.: Consequences of faster alignment of sequences. In: Automata, Languages, and Programming - 41st International Colloquium, ICALP 2014, Copenhagen, Denmark, 8–11 July 2014, Proceedings, Part I, pp. 39–51 (2014). https://doi.org/10.1007/978-3-662-43948-7_4
3. Artikis, A., Margara, A., Ugarte, M., Vansummeren, S., Weidlich, M.: Complex event recognition languages: Tutorial. In: Proceedings of the 11th ACM International Conference on Distributed and Event-based Systems, DEBS 2017, Barcelona, Spain, 19–23 June 2017, pp. 7–10 (2017). https://doi.org/10.1145/3093742.3095106
4. Baeza-Yates, R.A.: Searching subsequences. Theor. Comput. Sci. **78**(2), 363–376 (1991). https://doi.org/10.1016/0304-3975(91)90358-9
5. Barker, L., Fleischmann, P., Harwardt, K., Manea, F., Nowotka, D.: Scattered factor-universality of words. In: Jonoska, N., Savchuk, D. (eds.) DLT 2020. LNCS, vol. 12086, pp. 14–28. Springer, Cham (2020). https://doi.org/10.1007/978-3-030-48516-0_2
6. Bathie, G., Starikovskaya, T.: Property testing of regular languages with applications to streaming property testing of visibly pushdown languages. In: ICALP, LIPIcs, vol. 198, pp. 1–17. Schloss Dagstuhl - Leibniz-Zentrum für Informatik (2021)
7. Baumann, P., Ganardi, M., Thinniyam, R.S., Zetzsche, G.: Existential definability over the subword ordering. In: Berenbrink, P., Monmege, B. (eds.) Proceedings STACS 2022, LIPIcs, vol. 219, pp. 1–15 (2022)
8. Bespamyatnikh, S., Segal, M.: Enumerating longest increasing subsequences and patience sorting. Inf. Process. Lett. **76**(1), 7–11 (2000). https://doi.org/10.1016/s0020-0190(00)00124-1
9. Biedl, T.C.: Rollercoasters: long sequences without short runs. SIAM J. Discret. Math. **33**(2), 845–861 (2019). https://doi.org/10.1137/18M1192226
10. Bringmann, K.: Why walking the dog takes time: Frechet distance has no strongly subquadratic algorithms unless SETH fails. In: 55th IEEE Annual Symposium on Foundations of Computer Science, FOCS 2014, Philadelphia, PA, USA, 18–21 Oct 2014, pp. 661–670 (2014). https://doi.org/10.1109/FOCS.2014.76

11. Bringmann, K.: Fine-grained complexity theory (tutorial). In: 36th International Symposium on Theoretical Aspects of Computer Science, STACS 2019, 13–16 Mar 2019, Berlin, Germany. pp. 1–7 (2019). https://doi.org/10.4230/LIPIcs.STACS. 2019.4

12. Bringmann, K., Chaudhury, B.R.: Sketching, streaming, and fine-grained complexity of (weighted) LCS. In: Proceedings FSTTCS 2018, LIPIcs, vol. 122, pp. 1–16 (2018)

13. Bringmann, K., Künnemann, M.: Multivariate fine-grained complexity of longest common subsequence. In: Proceedings SODA 2018, pp. 1216–1235 (2018). https://doi.org/10.1137/1.9781611975031.79

14. Buss, S., Soltys, M.: Unshuffling a square is NP-hard. J. Comput. Syst. Sci. **80**(4), 766–776 (2014). https://doi.org/10.1016/j.jcss.2013.11.002

15. Crochemore, M., Hancart, C., Lecroq, T.: Algorithms on strings. Cambridge University Press (2007). https://doi.org/10.1017/cbo9780511546853

16. Crochemore, M., Melichar, B., Tronícek, Z.: Directed acyclic subsequence graph – overview. J. Discrete Algorithms **1**(3–4), 255–280 (2003). https://doi.org/10.1016/s1570-8667(03)00029-7

17. Crochemore, M., Porat, E.: Fast computation of a longest increasing subsequence and application. Inf. Comput. **208**(9), 1054–1059 (2010). https://doi.org/10.1016/j.ic.2010.04.003

18. Day, J.D., Fleischmann, P., Kosche, M., Koß, T., Manea, F., Siemer, S.: The edit distance to k-subsequence universality. In: Bläser, M., Monmege, B. (eds.) 38th International Symposium on Theoretical Aspects of Computer Science, STACS 2021, 16–19 Mar 2021, Saarbrücken, Germany (Virtual Conference). LIPIcs, vol. 187, pp. 1–19. Schloss Dagstuhl - Leibniz-Zentrum für Informatik (2021)

19. Day, J.D., Kosche, M., Manea, F., Schmid, M.L.: Subsequences with gap constraints: complexity bounds for matching and analysis problems. arXiv preprint arXiv:2206.13896 (2022)

20. Dilworth, R.P.: A decomposition theorem for partially ordered sets. Ann. Math. **51**(1), 161–166 (1950). https://doi.org/10.1007/978-1-4899-3558-8_1

21. Dudek, B., Gawrychowski, P., Gourdel, G., Starikovskaya, T.: Streaming regular expression membership and pattern matching. In: SODA, pp. 670–694. SIAM (2022). https://doi.org/10.1137/1.9781611977073.30

22. Dumitran, M., Gawrychowski, P., Manea, F.: Longest gapped repeats and palindromes. Discret. Math. Theor. Comput. Sci. **19**(4) (2017)

23. Erdős, P., Szekeres, G.: A combinatorial problem in geometry. Compos. Math. **2**, 463–470 (1935)

24. Fischer, J., Gawrychowski, P.: Alphabet-dependent string searching with wexponential search trees. In: Cicalese, F., Porat, E., Vaccaro, U. (eds.) CPM 2015. LNCS, vol. 9133, pp. 160–171. Springer, Cham (2015). https://doi.org/10.1007/978-3-319-19929-0_14

25. Fleischer, L., Kufleitner, M.: Testing Simon's congruence. In: Proceedings MFCS 2018. LIPIcs, vol. 117, pp. 1–13 (2018)

26. Fleischmann, P., Haschke, L., Huch, A., Mayrock, A., Nowotka, D.: Nearly k-universal words - investigating a part of simon's congruence. arXiv preprint arXiv:2202.07981 (2022)

27. Fredman, M.L.: On computing the length of longest increasing subsequences. Discret. Math. **11**(1), 29–35 (1975). https://doi.org/10.1016/0012-365x(75)90103-x

28. Freydenberger, D.D., Gawrychowski, P., Karhumäki, J., Manea, F., Rytter, W.: Testing k-binomial equivalence. In: Multidisciplinary Creativity, a collection of papers dedicated to G. Păun 65th birthday, pp. 239–248 (2015). arXiv preprint arXiv:1509.00622

29. Ganardi, M.: Language recognition in the sliding window model, Ph. D. thesis, University of Siegen, Germany (2019)

30. Ganardi, M., Hucke, D., König, D., Lohrey, M., Mamouras, K.: Automata theory on sliding windows. In: STACS, LIPIcs, vol. 96, pp. 1–14. Schloss Dagstuhl - Leibniz-Zentrum für Informatik (2018)

31. Ganardi, M., Hucke, D., Lohrey, M.: Querying regular languages over sliding windows. In: FSTTCS, LIPIcs, vol. 65, pp. 1–14. Schloss Dagstuhl - Leibniz-Zentrum für Informatik (2016)

32. Ganardi, M., Hucke, D., Lohrey, M.: Randomized sliding window algorithms for regular languages. In: ICALP, LIPIcs, vol. 107, pp. 1–13. Schloss Dagstuhl - Leibniz-Zentrum für Informatik (2018)

33. Ganardi, M., Hucke, D., Lohrey, M.: Sliding window algorithms for regular languages. In: Klein, S.T., Martín-Vide, C., Shapira, D. (eds.) LATA 2018. LNCS, vol. 10792, pp. 26–35. Springer, Cham (2018). https://doi.org/10.1007/978-3-319-77313-1_2

34. Ganardi, M., Hucke, D., Lohrey, M., Starikovskaya, T.: Sliding window property testing for regular languages. In: ISAAC, LIPIcs, vol. 149, pp. 1–13. Schloss Dagstuhl - Leibniz-Zentrum für Informatik (2019)

35. Garel, E.: Minimal separators of two words. In: Apostolico, A., Crochemore, M., Galil, Z., Manber, U. (eds.) CPM 1993. LNCS, vol. 684, pp. 35–53. Springer, Heidelberg (1993). https://doi.org/10.1007/BFb0029795

36. Gawrychowski, P., Kosche, M., Koß, T., Manea, F., Siemer, S.: Efficiently testing Simon's congruence. In: 38th International Symposium on Theoretical Aspects of Computer Science, STACS 2021, 16–19 Mar 2021, Saarbrücken, Germany (Virtual Conference), pp. 1–18 (2021). https://doi.org/10.4230/LIPIcs.STACS.2021.34

37. Gawrychowski, P., Manea, F., Serafin, R.: Fast and Longest Rollercoasters. Algorithmica 84(4), 1081–1106 (2022). https://doi.org/10.1007/s00453-021-00908-6

38. Giatrakos, N., Alevizos, E., Artikis, A., Deligiannakis, A., Garofalakis, M.: Complex event recognition in the big data era: a survey. VLDB J. 29(1), 313–352 (2019). https://doi.org/10.1007/s00778-019-00557-w

39. Halfon, S., Schnoebelen, P., Zetzsche, G.: Decidability, complexity, and expressiveness of first-order logic over the subword ordering. In: Proceedings LICS 2017, pp. 1–12 (2017). https://doi.org/10.1109/lics.2017.8005141

40. Hebrard, J.J.: An algorithm for distinguishing efficiently bit-strings by their subsequences. Theor. Comput. Sci. 82(1), 35–49 (1991)

41. Hegedüs, L., Nagy, B.: On periodic properties of circular words. Discrete Math. 339, 1189–1197 (2016). https://doi.org/10.1016/j.disc.2015.10.043

42. Hunt, J.W., Szymanski, T.G.: A fast algorithm for computing longest common subsequences. Commun. ACM 20(5), 350–353 (1977). https://doi.org/10.1145/350581.359603

43. Impagliazzo, R., Paturi, R.: On the complexity of k-SAT. J. Comput. Syst. Sci. 62(2), 367–375 (2001). https://doi.org/10.1006/jcss.2000.1727

44. Impagliazzo, R., Paturi, R., Zane, F.: Which problems have strongly exponential complexity? J. Comput. Syst. Sci. 63(4), 512–530 (2001). https://doi.org/10.1006/jcss.2001.1774

45. Karandikar, P., Kufleitner, M., Schnoebelen, P.: On the index of Simon's congruence for piecewise testability. Inf. Process. Lett. **115**(4), 515–519 (2015). https://doi.org/10.1016/j.ipl.2014.11.008
46. Karandikar, P., Schnoebelen, P.: The height of piecewise-testable languages with applications in logical complexity. In: Proceedings CSL 2016, LIPIcs, vol. 62, pp. 1–22 (2016)
47. Karandikar, P., Schnoebelen, P.: The height of piecewise-testable languages and the complexity of the logic of subwords. Log. Methods Comput. Sci. **15**(2) (2019)
48. Kärkkäinen, J., Sanders, P., Burkhardt, S.: Linear work suffix array construction. J. ACM **53**(6), 918–936 (2006). https://doi.org/10.1145/1217856.1217858
49. Kosche, M., Koß, T., Manea, F., Pak, V.: Subsequences in bounded ranges: matching and analysis problems. arXiv preprint arXiv:2207.09201 (2022)
50. Kosche, M., Koß, T., Manea, F., Siemer, S.: Absent subsequences in words. In: Bell, P.C., Totzke, P., Potapov, I. (eds.) RP 2021. LNCS, vol. 13035, pp. 115–131. Springer, Cham (2021). https://doi.org/10.1007/978-3-030-89716-1_8
51. Kuske, D.: The subtrace order and counting first-order logic. In: Fernau, H. (ed.) CSR 2020. LNCS, vol. 12159, pp. 289–302. Springer, Cham (2020). https://doi.org/10.1007/978-3-030-50026-9_21
52. Kuske, D., Zetzsche, G.: Languages ordered by the subword order. In: Bojańczyk, M., Simpson, A. (eds.) FoSSaCS 2019. LNCS, vol. 11425, pp. 348–364. Springer, Cham (2019). https://doi.org/10.1007/978-3-030-17127-8_20
53. Lejeune, M., Leroy, J., Rigo, M.: Computing the k-binomial complexity of the Thue–Morse word. In: Hofman, P., Skrzypczak, M. (eds.) DLT 2019. LNCS, vol. 11647, pp. 278–291. Springer, Cham (2019). https://doi.org/10.1007/978-3-030-24886-4_21
54. Leroy, J., Rigo, M., Stipulanti, M.: Generalized Pascal triangle for binomial coefficients of words. Electron. J. Combin. **24**, 36–44 (2017)
55. Lokshtanov, D., Marx, D., Saurabh, S.: Lower bounds based on the exponential time hypothesis. Bull. EATCS **105**, 41–72 (2011). http://eatcs.org/beatcs/index.php/beatcs/article/view/92
56. Maier, D.: The complexity of some problems on subsequences and supersequences. J. ACM **25**(2), 322–336 (1978). https://doi.org/10.1145/322063.322075
57. Manea, F., Tiseanu, C.: The hardness of counting full words compatible with partial words. J. Comput. Syst. Sci. **79**(1), 7–22 (2013). https://doi.org/10.1016/j.jcss.2012.04.001
58. Mateescu, A., Salomaa, A., Yu, S.: Subword histories and Parikh matrices. J. Comput. Syst. Sci. **68**(1), 1–21 (2004). https://doi.org/10.1016/j.jcss.2003.04.001
59. Pin, J.-E.: The consequences of Imre simon's work in the theory of automata, languages, and semigroups. In: Farach-Colton, M. (ed.) LATIN 2004. LNCS, vol. 2976, p. 5. Springer, Heidelberg (2004). https://doi.org/10.1007/978-3-540-24698-5_4
60. Pin, J.É.: The influence of Imre Simon's work in the theory of automata, languages and semigroups. Semigroup Forum **98**(1), 1–8 (2019). https://doi.org/10.1007/s00233-019-09999-8
61. Riddle, W.E.: An approach to software system modelling and analysis. Comput. Lang. **4**(1), 49–66 (1979). https://doi.org/10.1016/0096-0551(79)90009-2
62. Rigo, M., Salimov, P.: Another generalization of abelian equivalence: binomial complexity of infinite words. Theor. Comput. Sci. **601**, 47–57 (2015). https://doi.org/10.1016/j.tcs.2015.07.025

63. Salomaa, A.: Connections between subwords and certain matrix mappings. Theoret. Comput. Sci. **340**(2), 188–203 (2005). https://doi.org/10.1016/j.tcs.2005.03.024

64. Seki, S.: Absoluteness of subword inequality is undecidable. Theor. Comput. Sci. **418**, 116–120 (2012). https://doi.org/10.1016/j.tcs.2011.10.017

65. Shaw, A.C.: Software descriptions with flow expressions. IEEE Trans. Software Eng. **4**(3), 242–254 (1978). https://doi.org/10.1109/TSE.1978.231501

66. Simon, I.: Hierarchies of events with dot-depth one, Ph. D. thesis, University of Waterloo (1972)

67. Simon, I.: Piecewise testable events. In: Brakhage, H. (ed.) GI-Fachtagung 1975. LNCS, vol. 33, pp. 214–222. Springer, Heidelberg (1975). https://doi.org/10.1007/3-540-07407-4_23

68. Simon, I.: Words distinguished by their subwords (extended abstract). In: Proceedings WORDS 2003. TUCS General Publication, vol. 27, pp. 6–13 (2003)

69. Tronicek, Z.: Common subsequence automaton. In: Champarnaud, J.-M., Maurel, D. (eds.) CIAA 2002. LNCS, vol. 2608, pp. 270–275. Springer, Heidelberg (2003). https://doi.org/10.1007/3-540-44977-9_28

70. Williams, V.V.: Hardness of easy problems: basing hardness on popular conjectures such as the strong exponential time hypothesis (invited talk). In: 10th International Symposium on Parameterized and Exact Computation, IPEC 2015, 16–18 Sept 2015, Patras, Greece, pp. 17–29 (2015). https://doi.org/10.4230/LIPIcs.IPEC.2015.17

71. Zetzsche, G.: The complexity of downward closure comparisons. In: Proceedings ICALP 2016. LIPIcs, vol. 55, pp. 1–14 (2016)

72. Zhang, H., Diao, Y., Immerman, N.: On complexity and optimization of expensive queries in complex event processing. In: International Conference on Management of Data, SIGMOD 2014, Snowbird, UT, USA, 22–27 June 2014, pp. 217–228 (2014). https://doi.org/10.1145/2588555.2593671

73. Kleest-Meißner, S., Sattler, R., Schmid, M.L., Schweikardt, N., Weidlich, M.: Discovering event queries from traces: laying foundations for subsequence-queries with wildcards and gap-size constraints. In: Olteanu, D., Vortmeier, N. (eds.) 25th International Conference on Database Theory, ICDT 2022, 29 March to 1 April 2022, Edinburgh, UK (Virtual Conference). LIPIcs, vol. 220, pp. 1–21. Schloss Dagstuhl - Leibniz-Zentrum für Informatik, Dagstuhl (2022). https://doi.org/10.4230/LIPIcs.ICDT.2022.18

Canonization of Reconfigurable PT Nets in Maude

Lorenzo Capra[✉]

Dipartimento di Informatica, Università degli Studi di Milano, Milan, Italy
capra@di.unimi.it

Abstract. Petri Nets are a central model for concurrent or distributed systems but are not expressive enough to specify a system's dynamic reconfiguration. Rewriting Logic, in turn, has proved to be a suitable framework for several formal models of distributed systems. We have recently proposed an efficient Maude formalization of dynamically reconfigurable PT nets. In this paper, we address the scalability of this model using a canonization technique for PT systems integrated into Maude.

1 Introduction

Several types of modern SW/HW systems, among which those distributed, embedded, self-adapting, and automated, operate under varying conditions in highly dynamic environments. System components may become temporarily or permanently unavailable, may appear/disappear, e.g., due to failures or a dynamic load balancing. These systems use dynamic-reconfiguration procedures that overlap with the system's functionality. There is an impelling need for formal methods by which we can specify both the system's base layout and reconfiguration to assess design choices and verify the system's behaviour at run-time.

Petri nets (PNs) are a central model of concurrent or distributed systems, but they lack the flexibility to specify dynamically reconfigurable systems. PN extensions have been proposed in which enhanced expressivity is not adequately supported by automated analysis techniques. The most significant representatives meet the "nets within nets" paradigm, introduced in [20], and are High-Level PNs enriched with algebraic annotations for the manipulation of net-tokens, e.g., [4,10,12]. Reconfigurable PN should be mentioned as well, a family of PN-based formalisms composed of a classical marked PN and a separated set of rewrite rules specified as *pushouts*, according to algebraic Graph Transformations Systems, [9,11,13,17]. Most research in this field has focused on trying to formulate these models as $\mathcal{M}$-adhesive categories. See [15] for a survey.

We have recently proposed [6,7] a Maude [8] formalization of "rewritable" Place-Transition (PT) nets with inhibitor arcs (a Turing-complete formalism), somehow inspired to Reconfigurable PN. Maude is a well-supported, purely declarative language with a sound concurrent semantics in rewriting-logic [3,14]. Compared to the Maude formalization of reconfigurable PNs given in [1,16] (for model-checking purposes), our encoding provides more data abstraction (to ease

A. W. Lin et al. (Eds.): RP 2022, LNCS 13608, pp. 160–177, 2022.
https://doi.org/10.1007/978-3-031-19135-0_11

the modeller task), is more compact (and then more efficient), and permits the definition of rewrite rules without the restrictions imposed by the pushout pattern.

In this paper, we address the scalability of our model. As usual in Graph Transformation Systems, the State Transition System associated with a `Maude` term representing a (rewritable) PT system should be defined up to (PT system) isomorphism. The solution we describe and discuss in this paper is an efficient (though upgradable) *canonization* procedure for PT systems fully integrated into `Maude` through a rich set of algebraic operators.

We use the same simple, tricky benchmark as in [6,7], a fault-tolerant production line of a Manufacturing System (MS). We provide some experimental evidence of the effectiveness (space reduction) of canonization and the induced overhead by formally verifying simple liveness properties with a base `Maude` tool.

2 Background: PT Nets and `Maude`

2.1 Place-Transition (PT) Nets with Inhibitor Arcs

A *multiset* (or *bag*) b on D is a map $b : D \to \mathbb{N}$. We say that $b(d)$ is the *multiplicity* of d and $d \in b$ if and only if $b(d) > 0$. The extension of basic arithmetic/relational operations on bags is component-wise. Let $Bag[D]$ denote the set of bags on D.

A PT *net* [18] is a 5-tuple (P, T, I, O, H), where: P, T are non-empty, finite sets such that $P \cap T = \emptyset$ and I, O, H are maps $T \to Bag[P]$. P and T hold the net *places* and *transitions*. The former – drawn as circles – represent system state variables, whereas the latter – drawn as bars – represent events causing local state changes. A distributed state of a PT net, called *marking*, is a bag $m \in Bag[P]$. A net is a kind of directed, bipartite multi-graph whose nodes are $P \cup T$. Maps I, O, H describe the *input*, *output*, and *inhibitor* edges, respectively (◯——▭   ▭——▶◯   ◯——◖). Let $f \in \{I, O, H\}$: if $k = f(t)(p) > 0$, then a weight-k edge of corresponding type links p to t.

The dynamics of a PT net is specified by the *firing rule*. A transition $t \in T$ is *enabled* in m if and only if: $I(t) \leq m \wedge H(t) >' m$ ($>'$ is the restriction of $>$ on the elements of $H(t)$). If t is enabled (in m) it may fire leading to $m' = m + O(t) - I(t)$. We use the notation: $m[t\rangle m'$.

A PT-*system* is a pair (N, m), where N is a net and m is a marking of N. The interleaving semantics of (N, m_0), where m_0 denotes the initial state, is specified by the *reachability graph* (RG), an edge-labelled directed graph (V, E) defined inductively: $m_0 \in V$; if $m \in V$ and $m[t\rangle m'$ then $m' \in V$, $m \xrightarrow{t} m' \in E$.

2.2 Rewriting Logic and the `Maude` System

`Maude` [8] is an expressive, purely declarative language with a rewriting logic semantics [3]. Statements are (conditional) *equations* (`eq`) and *rules* (`rl`). Both sides of a rule/equation are terms of a given *kind* that may contain variables. Rules and equations have a simple rewriting semantics in which instances of the lefthand side are replaced by corresponding instances of the righthand side. Maude's expressivity is achieved through: equational pattern matching modulo equational attributes, sub-typing and partiality, generic types, reflection.

A Maude *functional* module (fmod) contains only *equations* and (with all the imported modules) specifies an *equational theory* in membership equational logic [2]. Formally, such a theory is a pair $(\Sigma, E \cup A)$, where Σ is the signature, that is, the specification of all the (sub)sort, kind[1], and operator declarations; E is the set of (conditional) equations and membership axioms, and A is the set of operator equational attributes (assoc, comm, ..). The model of $(\Sigma, E \cup A)$ is the *initial algebra* (denoted $T_{\Sigma/E\cup A}$), which is both junk- and confusion-free and mathematically corresponds to the quotient of the (ground) term-algebra. Under certain conditions on E and A, the final values (*canonical* forms) of all ground terms form an algebra isomorphic to the initial-algebra.

A Maude *system module* (mod) contains *rewrite rules* and possibly equations. Rules represent local transitions in a concurrent system. Formally, a system module specifies a generalized *rewrite theory* [3], a four-tuple $\mathcal{R} = (\Sigma, E \cup A, \phi, R)$ where $(\Sigma, E \cup A)$ is a membership equational theory; R is a set of rewrite rules; ϕ specifies the operator arguments not touched by rules. A rewrite theory specifies a concurrent system. $(\Sigma, E \cup A)$ defines the algebraic structure of the states. R and ϕ specify the system's concurrent transitions. The initial model of $\mathcal{R}$ associates to each kind k a labeled transition system whose states are $T_{\Sigma/E\cup A,k}$, and whose transitions take the form: $[t] \overset{[\alpha]}{\to} [t']$, with $[t], [t'] \in T_{\Sigma/E\cup A,k}$, and $[\alpha]$ an equivalence class of rewrites modulo the equational theory of proof-equivalence. The executability condition for system modules is the *ground coherence*, which ensures that a rewriting strategy in which terms are first reduced to the canonical form then rewritten according to the rules is both sound and complete.

3   The Running Example: A Self-healing Production Line

We consider a simple manufacturing system (MS) with two symmetric production lines as an example (Fig. 1, top). This scenario has been used as a case study for Rewritable PT Nets in previous work [6], to which we refer for a detailed description. We will use the original place and transition names where possible to emphasise the similarities.

In the scenario we have raw material and two *production lines* (i.e. robots) t_1 and t_2 working on pieces of those, both performing the same job. The MS finally assembles pairs of worked pieces. During the execution one of the two lines may get faulty (a double failure has a negligible probability and is not modelled). In this case, the MS will adapt itself to preserve functionality (and worked pieces) using the available line. The bottom of Fig. 1 illustrates the adapted MS layout.

Upon another fault which affects the left line, the MS goes back to its nominal configuration after an hypothetical global repair. The parameter $M \in \mathbb{N}^+$ defines the number of raw pieces $(2 \cdot M)$ worked during an production cycle.

---

[1] A *kind* is an equivalence class grouping sorts directly or indirectly related by subsort order; terms in a kind without a specific sort are *undefined* or *error* terms.

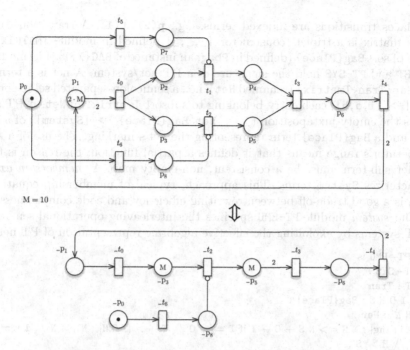

Fig. 1. The MS and its reconfiguration upon a fault on a line.

4 Encoding Rewritable PT Systems in Maude

The formalization relies on three generic functional modules, BAG{X}, MAP+{X,Y}, SET+{X} (the last two extensions of built-in modules). These modules may be arbitrarily nested thanks to a flexible mechanism of parameterized views (instantiating the type-parameters of a generic module). Differently from other Maude formalizations of PNs [16,19], bags are not merely represented as free commutative monoids on sets. A few bag-operators provide much more abstraction: _·_, _+_, _[_] _-_, _<=_, _>'_, set, _*_. The first two are *constructors*, i.e., appear in canonical forms. We can thus intuitively/conveniently represent a bag as a commutative/associative weighted sum, e.g., 3 . a + 1 . b. The module MAP+ defines a map term as a "set" of entries built using the associative/commutative juxtaposition _,_. Sub-sort Entry of Map has as a unique constructor _|->_. Module MAP+ supplies, among others, a predicate verifying the uniqueness of map's keys which is widely exploited (in data structures building on MAP) in membership equations implementing consistency checks.

PT System Formalization. The Maude specification in [6], here summarized, supplies an efficient operational semantics for *dynamically reconfigurable* PT nets and represents the basis for formalization. According to definition, however, dynamic adaptation comes down to net-tokens manipulation. Reconfiguration at the system-net level is part of ongoing work.

Places/transitions are indexed terms, e.g., p(2), t(1). A transition's incidence matrix is a triplet (constructor [_,_,_] defined in module IMATRIX) of terms of sort Bag{Place} (defined in Pbag, an instance of BAG{X})[2]. The modules PT-NET and PT-SYS hold the signature of a PT net/system. A net is a term of sort Map{Tran,Imatrix} (renamed Net), i.e., a semicolon-separated set of entries t(k)|-> [i,o,h], each entry belonging to subsort ImatrixT of Net. A PT system is the empty juxtaposition (__ : Net Bag{Place} -> [System]) of a Net term and a Bag{Place} term representing the net's marking. The use of a *kind* as operator's range means that it defines a partial function: the reason is that the net sub-term must be a consistent, non-empty map. A *membership axiom* characterizes System terms. This approach, typical of membership equational logic, is a good trade-off between rewriting efficiency and code compactness.

The *system* module PT-EMU specifies the interleaving operational semantics of PT systems by exploiting the effective algebraic representation of PT nets.

```
mod PT-EMU is
 pr PT-SYS .
 var T : Tran .
 vars I O H S : Bag{Place} .
 var N N' : Net .
 crl [firing] : N S => N S + O - I if I =/= O /\ T |-> [I,O,H] ; N' := N /\ I <= S
    /\ H >' S .
endm
```

The conditional rewrite rule firing intuitively encodes the PT firing rule. All the involved operators are bag-operators. The matching equation (t := t') in rule's condition makes it very compact. The model-specific part consists of a system module importing PT-EMU and containing two zero-arity operators of range Net and Bag{Place}, respectively, describing a given PT system.

Listing 1.1. Maude specification of self-healing MS

```
mod RWPT-FMS is
 protecting PT-RWLIB .
 op net : -> Net .
 op m0 : -> Pbag .
 op M : -> Nat . *** model's parameter (number of worked pieces)
 vars N N' : Net .
 vars TL TF : Tran . vars P2 P3 P4 P5 PF : Place .
 var S : Pbag .
 var K : NzNat .
 eq M = 50 .
 eq net = t(0) |-> [2 . p(1),1 . p(2) + 1 . p(3),nilP], t(1) |-> [1 . p(2),1 . p(4), 1 .
    p(7)], t(2) |-> [1 . p(3),1 . p(5),1 . p(8)], t(3) |-> [1 . p(4) + 1 . p(5),1 . p(6),
    nilP], t(4) |-> [1 . p(6),2 . p(1), nilP],t(5) |-> [1 . p(0),1 . p(7),nilP], t(6)
    |-> [1 . p(0), 1 . p(8),nilP].
 eq m0 = 2 * M . p(1) + 1 . p(0) .
```

---

[2] They represent the input, output, inhibitor connections, respectively.

crl [r1] : (N,t(0) |−> [2 . p(1),1 . P2 + 1 . P3,nilP],t(3) |−> [1 . P4 + 1 . P5,1 . p(6)
    ,nilP],TF |−> [1 . p(0),1 . PF,nilP],TL |−> [1 . P3,1 . P5,1 . PF]) S + 1 . PF =>
    (N,t(0) |−> [1 . p(1),1 . P2,nilP], t(3) |−> [2 . P4,1 . p(6),nilP]) set(S, P3,0)
    + S[P3] . P2 + 1 . p(0) if S[P5] = 0 .
crl [r2] : N S => net S + 1 . p(0) + M . P3 − M . P2 − 1 . p(7) − 1 . p(8) if 1 . P2 :=
    Out(N, t(0)) /\ 1 . P2 + 1 . P3 := Out(net, t(0)) /\ dead(N S) .
endm

4.1   Reachability Properties of Rewritable PT Systems

Property 1 (correspondence between PT systems and well-defined terms). A PT
system $S = (N, m)$ has an associated ground term of sort System, vice-versa, a
ground term of sort System represents a PT system (up to isomorphism).

Let r be a rewrite rule, t, t' two ground terms of kind k. The notation $t \overset{r(\sigma)}{\to} t'$
means that 1) the rule's lefthand side $u \in T_{\Sigma(X),k}$ matches t (i.e., there is a
ground substitution σ such that $\sigma(u) = t)^3$, 2) t is rewritten to t' using (r,σ).

Given a PT system $S = (N, m_0)$, let RWPT-S represent a system module in
which the term (net m0) encodes S, R be the set of rewrite rules defined in
RWPT-S, and S-EMU a system module importing both PT-EMU and RWPT-S. The
interleaving semantics of the rewritable PT system specified by RWPT-S is:

Definition 1 (State-transition system of RWPT-S). *Let* $R' = R \cup \{\text{firing}\}$.
$RWLT_S$ is an edge-labelled, directed graph (V_{RW_S}, E_{RW_S}) *inductively defined:*
(net m0) $\in V_{RW_S}$; *if* $s \in V_{RW_S}$ *and* $s \overset{r(\sigma)}{\to} s'$ *then:* $s' \in V_{RW_S}$, $s \overset{r(\sigma)}{\to} s' \in E_{RW_S}$.

The Maude's command search explores the state-space of a term by executing
rewrite rules in a breadth-first way, coherently with the definition above. By the
way, RWPT-S includes the ordinary behaviour of the PT system S.

Property 2 (RG inclusion). $RWLT_S$ contains a sub-graph isomorphic to RG_S.

Notice that, in the event of badly defined/used rules, we may reach undefined
(error) states, despite (net m0) well-definiteness.

Definition 2 (Well-defined specification). *RWPT-S specifies a rewritable
PT system if and only if all reachable states in V_{RW_S} are terms of sort System.*

Rule Validation. The module RWPT-FMS is well-defined. In general, however,
ensuring the well-definiteness of a Maude specification of a rewritable PT may
not be simple. There are two possible approaches. One consists of defining struc-
turally valid rewrite rules and works also in the event of an infinite state-space.

Definition 3 (Valid rewrite rule). $r \in R$ *is valid if and only if, for any
ground term s of sort System, if $s \overset{r(\sigma)}{\to} s'$ then s' is of sort System.*

---

[3] σ is null is u is a ground term; if r is a conditional rule σ may involve free variables
introduced by matching equations used in the rule's condition.

We may indeed rephrase any rewrite rule crl [r] : s => s' if cond, where s, s' $\in T_{\Sigma(X),\text{System}}$, as a valid one using the built-in *sort membership*:

crl [vr] s => s' if cond $/\backslash$ s' :: System.

The weak spot of this elegant and efficient solution is that it may shadow bad design choices. Otherwise, we may define rules using exclusively *safe* net-operators. For example, the operator setw (defined in module PT-RWLIB) always results in a term of sort Net. The operator setwS, which builds on setw, guarantees that the resulting term is a non-empty PT net (note the use of owise equation attribute mixed with a matching equation).

5   Canonization of Rewritable PT

In this section, we illustrate through an example the procedure for canonizing (reconfigurable) PT systems. We mention the main Maude operators employed (most of which are collected in CAN-PT-SYS module) by omitting the technical details of their definition. We refer to unlabeled (uncoloured) PT nodes. We discuss the possible extension of the technique to coloured (i.e., statically partitioned) nodes and related (dis)advantages in the section on ongoing work.

We start by briefly presenting some related work on graph isomorphism/canonization and giving some basic notions used in the sequel.

5.1   Graph Isomorphism and Canonization: An Overview

PT System Isomorphism. An isomorphism ϕ between PT systems $S = (N, m)$, $S' = (N, m')$ is a pair of bijections $\phi_p : P \to P'$, $\phi_t : T \to T'$, preserving edge connections and marking. S and S' are isomorphic if and only there is an isomorphism mapping S to S'. If $N = N'$ and $T = T'$ then we speak of automorphism. In the sequel, we basically restrict to automorphisms. Two places (transitions) of S are *automorphic equivalent* (or *symmetric*) if and only if there is an automorphism mapping one to the other. Checking graph isomorphism is in general complex: this well-studied problem belongs to NP, but is thought to be neither P nor NP-complete. Canonization is an approach which consists in finding an (usually "minimal") form such that $S \cong S' \Leftrightarrow can(S) = can(S')$.

5.2   Base Notations/Definitions for PT Canonization (in Maude)

We denote with $(A, <)$, or $<_A$ (possibly omitting A if implicit) a strict *total order* on a set (sort, in Maude's parlance) A (or a totally ordered set/sort, depending on the context). Let $\mathcal{LIST}(A)$ be the set/sort of lists defined on $(A, <)$. Then $(\mathcal{LIST}(A), <)$ is the *lexicographic* order between lists induced by $<_A$. The canonization of PT systems operates on list of pairs. Given $(A_1, <)$, $(A_2, <)$, then $(A_1 \times A_2, <)$ is: $\langle a_1, a_2 \rangle < \langle a_1', a_2' \rangle$ iff $a_1 < a_1' \vee (a_1 = a_1' \wedge a_2 < a_2')$. On lists of (sortable) pairs we may use the total order(s) induced by pair components, further to the default lexicographic order. Let l $(l') \in \mathcal{LIST}(A_1 \times A_2)$: we denote with l_1 ($\in \mathcal{LIST}(A_1)$) ($l_2 \in \mathcal{LIST}(A_2)$) the projection of l on A_1 (A_2). For any two l, l', $(\mathcal{LIST}(A_1 \times A_2), <_1)$ $((\mathcal{LIST}(A_1 \times A_2), <_2))$ is: $l <_1 l' \Leftrightarrow l_1 < l'_1$ ($l <_2 l' \Leftrightarrow l_2 < l'_2$).

Ordering PT Systems in Maude. The (overloaded) operators $<$, $<_1$ are fomalized by simple equations. Specifically, (Place, $<$) and (Tran, $<$) map to (Nat, $<$), by considering node subscripts. For the rest, we exploit the library module SORTABLE-LIST{X :: STRICT-TOTAL-ORDER} by associating the module's parameter X (whose type is a *theory*) with a totally-ordered sort (of a concrete module). It supplies the operator op sort : List{X} -> List{X} which implements mergesort. (Pbag, $<$) maps to (List{Pair{Nat,Place}}, $<$). The operator op makeList : Pbag -> List{Pair{Nat,Place}} gets a list ouf of a multiset. Analogously, (Net, $<$) maps to (List{Pair{Tran,Imatrix}}, $<$) using op makeList : Net -> List{Pair{Tran,Imatrix}} to get a list out of a Net. The total order (Imatrix, $<$) is formalized (in Imatrix module) as:

```
vars X Y Z X' Y' Z' : Pbag .
op _<_ : Imatrix Imatrix -> Bool .
eq [X,Y,Z] < [X',Y',Z'] = X < X' or-else X == X' and-then (Y < Y' or-else Y == Y'
     and-then Z < Z').
```

Finally, we define (System, $<$) as:

```
op _<_ : System System -> Bool .
vars N N' : Net .
vars S S' : Pbag .
eq N S < N' S' = S < S' or-else S == S' and-then N < N' .
```

The *canonized form* of the PT system encoded by (net m) is the least System term isomorphic to (net m), according to (System, $<$).

Canonization. We illustrate the main steps of PT system canonization through our running example (Listing 1.1). We initially refer to the system term aliased by (net m0), with M = 1.

The algorithm uses (Pbag, $<_1$), namely, (List{Pair{Nat,Place}}, $<_1$), and (Imatrix, $<_1$), which is built on top of (Pbag, $<_1$) and is formalized by:

```
vars X Y Z X' Y' Z' : Pbag .
op _<1_ : Imatrix Imatrix -> Bool .
eq [X,Y,Z] <1 [X',Y',Z'] = X <1 X' or-else X == X' and-then (Y <1 Y' or-else Y
     == Y' and-then Z <1 Z').
```

Canonization is place-driven (we may consider transitions only at the end) and logically consists of two stages: first, we canonize the Pbag (marking) sub-term and then the Net sub-term. The order in which we consider these sub-terms matters: marking's canonization indeed affects net's canonization.

As anticipated, canonization is incremental and monotonic: we repeatedly permute pairs of places using op swap : System Place Place -> System[4] until we eventually reach the minimal form. Places are considered one by one in

---

[4] swap is overloaded, e.g., by op swap : Imatrix Place Place -> Imatrix.

increasing order. For each place, we search for possible candidates to swap in a restricted sub-list of greater by possibly pruning the swap sequences to try out. By the way, any sequence of swaps defines an automorphism.

As a preliminary key step, we get a sorted list of matrices out of the net sub-term (op makeAdjList : Net -> List{Imatrix}) using (Imatrix, $<_1$). This allows us to divide (op regroup : List{Imatrix} -> List{List{Imatrix}}) the list of matrices into contiguous blocks of "similar" elements composed of multisets having the same weights. We, therefore, have to consider these elements together to calculate the isomorphic minimal form. We prefix the list of blocks with the marking sub-term, to get the expected ordering effect. It is worth noting that the final form matches this disposal, up to reordering elements inside blocks (according to (Imatrix, $<$)). The luckiest situation occurs when blocks are singletons: in that case, no reordering is needed.

The sorted list of blocks we obtain from net m0 is as follows:

$1 \cdot p(0) + 2 \cdot p(1)$,	block 1 (marking)
$[1 \cdot p(0), 1 \cdot p(7), \text{nilP}], [1 \cdot p(0), 1 \cdot p(8), \text{nilP}]$,	block 2
$[1 \cdot p(2), 1 \cdot p(4), 1 \cdot p(7)], [1 \cdot p(3), 1 \cdot p(5), 1 \cdot p(8)]$,	block 3
$[1 \cdot p(6), 2 \cdot p(1), \text{nilP}]$,	block 4
$[1 \cdot p(4) + 1 \cdot p(5), 1 \cdot p(6), \text{nilP}]$,	block 5
$[2 \cdot p(1), 1 \cdot p(2) + 1 \cdot p(3), \text{nilP}]$.	block 6

For each place p(i), starting from the smallest one, we scan the sorted list of blocks by considering the input/output/inhibitor components in sequence. We search for candidates to swap with p(i), namely places {p(j)}, $j > i$, whose multiplicity is minimal and is less than or equal to that of p(i), if that is not zero. If this set is empty, we go on with the next place. If it is a singleton, we carry out the swap p(i) $\longleftrightarrow$ p(j). Otherwise, we may have to (recursively) branch the canonization process on sub-lists to determine the sequence of swaps leading to the minimal list. In the last two cases, we need to reorder non-singleton blocks, as explained. No backtracking on the computation tree is needed.

This procedure relies on the assumption (trivially met at the beginning) that at any moment, any permutation p(i) $\longleftrightarrow$ p(j), with $j < i$, results in a non-smaller list. The proof (omitted, being only technical stuff) is by contradiction.

Going back to the example, we easily argue that the marking (block 1) is the smallest multiset of that shape. Therefore, no swap involves p(0), p(1) and we pass considering p(2) by directly jumping to block 2. Looking at the block's output components, we figure out that there are two candidates to swap with p(2), namely, p(7), p(8). At this point the computation splits into two branches: let us call them b1 and b2, and denote sub-branches as b1.1, b1.2, etc. The final, minimal form results from (recursively) taking the minimum of local branches. After p(2) we have, necessarily, to swap p(3) with either p(7) or p(8). We get (we implicitly carry out a local reorder after any swap):

b1 $(p(2) \longleftrightarrow p(7); p(3) \longleftrightarrow p(8))$

$[1 \cdot p(0), 1 \cdot p(2), \texttt{nilP}], [1 \cdot p(0), 1 \cdot p(3), \texttt{nilP}],$	block 2
$[1 \cdot p(7), 1 \cdot p(4), 1 \cdot p(2)], [1 \cdot p(8), 1 \cdot p(5), 1 \cdot p(3)],$	block 3
...,	block 4
...,	block 5
$[2 \cdot p(1), 1 \cdot p(7) + 1 \cdot p(8), \texttt{nilP}]$.	block 6

b2 $(p(2) \longleftrightarrow p(8); p(3) \longleftrightarrow p(7))$

$[1 \cdot p(0), 1 \cdot p(2), \texttt{nilP}], [1 \cdot p(0), 1 \cdot p(3), \texttt{nilP}],$	block 2
$[1 \cdot p(7), 1 \cdot p(5), 1 \cdot p(2)], [1 \cdot p(8), 1 \cdot p(4), 1 \cdot p(3)],$	block 3
...,	block 4
...,	block 5
$[2 \cdot p(1), 1 \cdot p(7) + 1 \cdot p(8), \texttt{nilP}]$.	block 6

The next place to consider in each branch is $p(4)$. Since block 2 (which coincides in b1 and b2) now contains places lesser than $p(4)$, we can jump to block 3. We figure out that the candidates for swapping with $p(4)$ are (again) $p(7), p(8)$, in both the main branches. That results in sub-branches, as described below. As before, after the swap of $p(4)$, there is a mandatory sequence of swaps that lead to local, minimal forms. Let us follow the sub-paths from b1, one of which leads to the global minimum (we don't describe the sub-branches of b2). Notice that there is no further branching from that point on.

b1.1 $(p(4) \longleftrightarrow p(7); p(5) \longleftrightarrow p(8))$

$[1 \cdot p(4), 1 \cdot p(7), 1 \cdot p(2)], [1 \cdot p(5), 1 \cdot p(8), 1 \cdot p(3)],$	block 3
$[1 \cdot p(6), 2 \cdot p(1), \texttt{nilP}],$	block 4
$[1 \cdot p(7) + 1 \cdot p(8), 1 \cdot p(6), \texttt{nilP}],$	block 5
$[2 \cdot p(1), 1 \cdot p(4) + 1 \cdot p(5), \texttt{nilP}]$.	block 6

Then we analyse $p(6)$: looking at block 3 (output components), there are seemingly two possible swaps of $p(6)$ with $p(7)$ and $p(8)$, i.e., two further sub-branches. But a simple heuristic allows us to prune $p(6) \longleftrightarrow p(8)$: it would result indeed in a greater sub-list, independently on the tail blocks. The only eligible sequence is $p(6) \longleftrightarrow p(7); p(7) \longleftrightarrow p(8)$. We finally get:

$[1 \cdot p(4), 1 \cdot p(6), 1 \cdot p(2)], [1 \cdot p(5), 1 \cdot p(7), 1 \cdot p(3)],$	block 3
$[1 \cdot p(8), 2 \cdot p(1), \texttt{nilP}],$	block 4
$[1 \cdot p(6) + 1 \cdot p(7), 1 \cdot p(8), \texttt{nilP}],$	block 5
$[2 \cdot p(1), 1 \cdot p(4) + 1 \cdot p(5), \texttt{nilP}]$.	block 6

As per the other sub-branch, we can make analogous considerations. We get:

b1.2 $(p(4) \longleftrightarrow p(8); p(5) \longleftrightarrow p(7); p(6) \longleftrightarrow p(7); p(7) \longleftrightarrow p(8))$

$[1 \cdot p(4), 1 \cdot p(6), 1 \cdot p(3)], [1 \cdot p(5), 1 \cdot p(7), 1 \cdot p(2)],$	block 3
$[1 \cdot p(8), 2 \cdot p(1), \texttt{nilP}],$	block 4
$[1 \cdot p(6) + 1 \cdot p(7), 1 \cdot p(8), \texttt{nilP}],$	block 5
$[2 \cdot p(1), 1 \cdot p(4) + 1 \cdot p(5), \texttt{nilP}]$.	block 6

It turns out that the minimal list of matrices is the one we obtain at the end of sub-branch b1.1 (with the addition of blocks 1 and 2). At that point, we only need to rearrange transition indices accordingly.

In general, we may also have to rescale the node subscripts so that they lie in $0 \ldots |P| - 1$, and $0 \ldots |T| - 1$, respectively. This may be done at the beginning or the end of the canonization, indifferently. In our example, that is not due.

The canonical form of the PT system in Fig. 1 (top) corresponds to the PT automorphism $\{p_2 \to p_4, p_3 \to p_5, p_4 \to p_6, p_5 \to p_7, p_6 \to p_8, p_7 \to p_2, p_8 \to p_3\}$, $\{t_0 \to t_6, t_1 \to t_2, t_2 \to t_3, t_3 \to t_5, t_5 \to t_0, t_6 \to t_1\}$ (identities are implicit).

Here, we are not interested in studying the theoretical complexity of the canonization procedure. We supply a few experimental pieces of evidence in the next section. We may significantly improve the algorithm by implementing more sophisticated heuristics to prune branches in the computation tree (other than that used in b1.1 and b1.2). For example, it should even be possible to avoid the split of the two main branches by looking at the structure of the outgoing lists.

Nonetheless, we point out that compared to a brute-force approach enumerating the 9! permutations on places (more those on transitions), we have only carried out a few dozens of swaps. We believe that such an improvement is not occasional due to some regularity shown by most realistic models.

Implementing the PT canonization directly in Maude, however, is far from intuitive. The entire process is divided into several (recursive) sub-tasks, each matching an operator. Let us mention a few, among which the top operator.

```
op canonize : System −> System . *** top canonization operator
op canonizeNet : List{List{Imatrix}} List{Place} −> List{List{Imatrix}} .
op canonizeM : Pbag List{List{Imatrix}} List{Place} −> List{List{Imatrix}} .
op canonizeP : List{List{Imatrix}} Place −> List{List{Imatrix}} .
op candidates : List{List{Imatrix}} Place −> Set{Place} .
```

5.3   Equivalences and Rewrite Rules: Making Rules Symmetric

The canonization of System terms allows the automatic detection of behavioural equivalences (symmetries) that a PT system may exhibit during its evolution. Equivalences are of two types: those due to the inner dynamics of a PT system (expressed by the firing rule) and those due to the structural changes that a PT system undergoes (specified by the other rewrite rules).

Syntactically, the right-hand side of *any* rewrite rule needs to be surrounded by the top canonization operator. Thus, the firing rule becomes:

```
var T : Tran .
vars I O H S : Bag{Place} .
var N N' : Net .
crl [syfiring] : N S => canonize(N S + O − I) if I =/= O /\ T |−> [I,O,H] ; N' := N
    /\ I <= S /\ H >' S .
```

The *canonized state-transition system* of a rewritable PT system encoded by the System term (net m0) is the state transition system generated from the canonized form of net m0 (Definition 1), assuming that all rewrite rules of System type take the form crl [r] : s => canonize(s') if cond (as usual, we consider conditional rules as more general).

We may conveniently define an alias for the canonized, initial PT system.

```
op cansys0 : -> System .
eq cansys0 = canonize(net m0) .
```

Unfortunately, rules are not terms, so they are not affected by canonization. For the canonized state transition system to preserve the properties of the ordinary one, rewrite rules have to be *symmetric*.

Definition 4 (Symmetric rewrite rule). crl [r] : s => s' if cond *is symmetric if and only if only non ground terms occur both in* s *and* cond.

Since any term in our specification is made up of sub-terms of sorts Place and Tran, we may rephrase Definition 4 by saying that a rule r is symmetric if only *variable*-terms of these two sorts occur in r's left-hand side and condition.

The firing rule trivially meets Definition 4. But the two rules specifying the evolution of the MS upon a fault (Listing 1.1) do not. For example, the ground terms p(0),p(1),p(6) appear in $r1$'s left-hand side. In the listing below, we show a possible translation of these rules into a symmetric form. As per $r1$, we straightforwardly replace ground terms with corresponding variables. The translation of $r2$ is a bit trickier since the deadlock (triggering the rewriting of the system into its initial shape, while retaining worked pieces) factorizes two different states. It is worth noticing that (modulo the canonization) this translation *exactly* preserves the behaviour of the original rules. The use of variables instead of ground terms only causes a negligible overhead in rule matching.

Listing 1.2. symmetric rewrite rules of the self-healing MS

```
vars N N' Net .
vars T0 T1 T3 TL TF : Tran .
vars P1 P2 P3 P4 P5 P6 PF : Place .
var S : Pbag .
...
crl [syr1] : (N, T0 |-> [2 . P1, 1 . P2 + 1 . P3, nilP] , T3 |-> [1 . P4 + 1 . P5, 1 . P6
    , nilP],TF |-> [1 . P0,1 . PF,nilP],TL |-> [1 . P3,1 . P5, 1 . PF]) S + 1 . PF
 => canonize((N, T0 |-> [1 . P1, 1 . P2, nilP], T3 |-> [2 . P4, 1 . P6, nilP])
    set(S, P3, 0) + S[P3] . P2 + 1 . P0) if S[P5] = 0 .
crl [syr2] : N S => canonize(net 1 . p(0) + M . p(3) + sd(S[P2], M) . p(2) + if S[P3]
    == 1 then 1 . p(4) else nilP fi) if N', T0 |-> [1 . P1, 1 . P2, nilP], T1 |-> [1
    . P2, 1 . P3, 1 . P4]) := N /\ dead(N S) .
```

5.4 Canonization and Reachability

The following properties formalize that the canonized state transition system is a *quotient* of the original (non canonized) one. We denote with s, s', u, u' (final) ground terms of sort System, with $\hat{s}$ the canonized form of s, with r a rewrite rule $t \Rightarrow t'$ which meets Definition 4, and with $\hat{r}: t \Rightarrow canonize(t')$ (σ is a ground substitution of a rule's left term's variables, $u \cong s$ is the same as $\hat{u} = \hat{s}$).

Property 3 (correspondence between transitions). Let $s \xrightarrow{r(\sigma)} s'$. Then $\hat{s} \xrightarrow{\hat{r}(\phi(\sigma))} \hat{s}'$, where ϕ is the isomorphism from s to $\hat{s}$.

Property 4 (source correspondence). Let $s \xrightarrow{\hat{r}(\sigma)} s'$. Then $\forall u \cong s \; \exists u' \cong s'$ s.t. $u \xrightarrow{r(\phi(\sigma))} u'$, where ϕ is the isomorphism from s to u.

The analogous of Property 4 for the target node of a canonized state transition doesn't hold, because we admit the presence of ground terms on the right-hand side of a symmetric rule. For example, in $syr2$ we use the `net` ground term to refer to the initial structure of the MS.

Theoretically, this doesn't represent a restriction because we always reason "up to isomorphism". The use of ground terms on the right-hand side of rules permits, actually, an interesting optimization: If the outcome of a `system` transformation, under certain conditions, is known "a priori" then we can avoid repeating canonization using the simple pattern below.

```
op s : -> System .
eq s = canonize(...) . *** ... ground term to use on rule's right-hand side
crl [optr] t => s if c . *** t and c are terms satisfying Def. 1
```

We could use this pattern to improve the efficiency of our running example at the cost of splitting rule $syr2$ in two, given that (using PT structural analysis) we know how and when the system's reconfiguration takes place [6,7].

A similar but general optimization concerns the firing rule. Do we always need to canonize a `System` term after a transition's firing? In some cases, we can not do it. More precisely, whenever the change of marking preserves the marking canonization. For example, if the marking of a canonized system is $1 \cdot p_0 + 2 \cdot p_1$ and the new reached marking is $1 \cdot p_0 + 3 \cdot p_1$ then we can drop the canonization of the (new) `System` term. That would require splitting the corresponding rewrite rule by refining its condition and is part of our ongoing work.

Let us finally (briefly) discuss an important aspect of reachability. How many *different* instances do really represent a (canonized) `System` term and a canonized firing? We consider, once again, the PT system in Fig. 1 (top), whose `Maude` encoding is Listing 1.1 (term `net m0`) and which we have used to illustrate the canonization procedure. For simplicity, in the sequel, we mix the use of the classical PT notation and the algebraic one. We initially ignore canonization.

Consider the firing sequences 1) $m_0[t_0; t_1 > m_1$, with $m_1 = 1 \cdot p_0 + 1 \cdot p_3 + 1 \cdot p_4$, 2) $m_0[t_0; t_2 > m_2$, with $m_2 = 1 \cdot p_0 + 1 \cdot p_2 + 1 \cdot p_5$. The markings m_1, m_2 have the same minimal form $(1 \cdot p_0 + 1 \cdot p_1 + 1 \cdot p_2)$ and the corresponding PT systems (`System` terms) collapse into the same canonized form (not represented). In the canonized state-transition system the two firing sequences above collapse as well.

To calculate the cardinality of a (canonized) `System` term (transition firing) we need to partition the places of the `Net` into classes of *automorphic equivalent*. In the example, p_2, p_3 are automorphic-equivalent, as well as, p_4, p_5. Thus in

m_1, m_2 (and in their canonized form as well) two places belongs to two different classes of size 2, and one to a singleton class. Using simple combinatorics, that means that the (canonized) System term represents two different instances. An algebraic characterization of this point is part of our ongoing work.

6 PT Canonization and Verification: Does It Scale?

In this section, we provide some experimental evidence of the use of canonization to face the complexity of formal verification of models specified with rewritable PT systems. We use a base analysis tool of the Maude system. We initially refer to our running example and then consider a parallel extension with N replicas of the MS production line to asses if the canonization technique scales up adequately.

A Maude system module specifies a rewrite theory. Therefore, it provides an executable formal model of a distributed system. Under some executability conditions (met by our specification), one can check that a model satisfies some properties or obtain counterexamples. This common model-checking builds on the inline search command, which explores the state transition system of a rewritable PT system using a breadth-first strategy. If a system generates a big (or infinite) state space we may set some bounds on search or use abstractions by adding equations in our specification. Under the finite reachability assumption, we can efficiently model-check any linear time temporal logic (LTL) property of a system module. Since our object is to evaluate the effectiveness (in terms of space reduction) of canonization and the induced time overhead, we model-check two simple invariant properties through the search command.

A base liveness property of a dynamically reconfigurable system is deadlock freedom. A straightforward way to check it is to issue the command below, which searches for any *final* states of the PT system specified in Listing 1.1 starting from the nominal configuration of the MS production line. The second version of the command searches through the canonized state transition system (the rewrite rules are those specified in Sect. 5.3, Listing 1.2) – FMS-EMU and FMS-EMU-CAN are convenience modules importing all the other necessary modules.

```
search in FMS–EMU : net m0 =>! X : System  .
search in FMS–EMU–CAN : cansys0 =>! X : System  .
```

As expected, the search has no matches, meaning that the MS is deadlock-free (using the LTL module, we asses that net m0, or its canonized form, is a home-state, i.e., the state-transition system is strongly-connected).

Another interesting search concerns possible dead states inside the different configurations (the nominal one and the two symmetric, faulty ones) that the system enters during its evolution. The command to issue is (* means "in zero or more steps", dead is a predicate defined in module PT-SYS):

```
search in FMS—EMU : net m0 =>* X:System such that dead(X:System) .
search in FMS—EMU—CAN : cansys0 =>* X:System such that dead(X:System) .
```

The first search matches six states (for any value of parameter M), two for each system configuration. The second search, instead, has only three matches (one refers to the nominal MS, two to its reconfigured layout) due to canonization.

Table 1 reports some data about the performances as the system's parameter M (half of the number of pieces worked in a production cycle) varies. These data refer to an Intel Core i7-6700 equipped with 32 GB RAM. The execution time value is the average of the two commands we have run. The second and third columns of the table refer to the ordinary state transition system and the canonized one, respectively. We note that canonization approximately halves the state space (independently on M) per the MS layout, made up of two symmetric lines. The execution time shows an increasing overhead due to canonization (for $M = 50$, it takes around 5 h). Most of this is because the implementation of the algorithm exploits trivial heuristics. We firmly believe that integrating more structural ones could dramatically drop the price of canonization.

Nonetheless, canonization becomes crucial when considering systems with many replicated components. Table 2, e.g., measures the performances of search on a variant of the running example with a higher parallelism degree (using the same HW configuration). In this variant, N replicas of the reconfigurable production line shown in Fig. 1 work in parallel by "synchronizing" at the beginning of a production cycle (for $N = 1$ the model coincides with that in Fig. 1). The value of M in each replica is 2. This model is formalized in [5] (Fig. 11) in a compact, parametric way using Symmetric Nets, a subclass of High-Level PN featuring a structured syntax which outlines the behavioural symmetries of a system. It corresponds to the term: `Par(net m0, N, p(1), empty)`, where

op `Par : System NzNat Set{Place} Set{Tran} —> System` .

is a net-algebra operator which creates N copies of the specified `System` by resizing the node subscripts and merging the specified sets of nodes –modulo the subscript resizing (we omit further details on the `Maude` specification).

As N grows, exploiting the system's symmetries through canonization of terms becomes the only way to deal with the combinatorial complexity of models: To give an idea, for $N = 4$ there are around sixty million ordinary states against under a million and a half canonized ones. The execution time data for $N = 4$ confirms the scalability due to canonization. As a final remark, the performances of the `Maude` specification of the self-healing MS outperform those shown by the emulation-based, Symmetric Nets model [5] of the same example.

Table 1. Performance of search command as M (pieces) varies

M	(ord.) # states	time (s.)	(can.) # states	time (s.)
2	92	0.004	48	0.136
5	1232	0.15	604	1.73
10	8932	2.19	4398	29.65
20	94008	36.09	46904	357.60
50	2186132	862.53	1098644	17061

Table 2. Performance of `search` command as N (replicas) varies

N	(ord.)	# states	time (s.)	(can.)	# states	time (s.)
1		92	0.007		48	0.198
2		7618	1.81		1798	15.36
4		60639296	29136		1337404	26131

7   Conclusion, Open Issues and Ongoing Work

We have presented an efficient canonization technique for a class of "rewritable" PT systems recently formalized in `Maude`. This approach faces a major scalability issue of Graph Transformation Systems, namely, the automatic recognition of isomorphic graphs (PT systems). We have used a simple, tricky example throughout the paper. We have finally discussed the effectiveness and overhead of canonizing rewritable PT systems by reporting some experimental evidences of a base `Maude` model-checker. Even if the outcomes seem promising in terms of scalability, much work has to done to reduce the canonization overhead.

Open issues We have currently implemented a few trivial heuristics which help prune some branches of the canonization's computation tree in particularly simple conditions. More structured heuristics (e.g., avoiding the canonization of the `Net` sub-term after a PT transition's firing which retains the marking canonization) may dramatically reduce the canonization's inefficiency in most practical cases. Two more systematic optimizations, currently under study, are the use of labelling to partition PT nodes in classes of automorphic equivalent and the exploitation of net-algebra operators to specify bunches of symmetric system components (related in some way).

References

1. Barbosa, P., et al.: SysVeritas: a framework for verifying IOPT nets and execution semantics within embedded systems design. In: Camarinha-Matos, L.M. (ed.) DoCEIS 2011. IAICT, vol. 349, pp. 256–265. Springer, Heidelberg (2011). https://doi.org/10.1007/978-3-642-19170-1_28
2. Bouhoula, A., Jouannaud, J.P., Meseguer, J.: Specification and proof in membership equational logic. Theor. Comput. Sci. **236**(1), 35–132 (2000). https://doi.org/10.1016/S0304-3975(99)00206-6
3. Bruni, R., Meseguer, J.: Generalized rewrite theories. In: Baeten, J.C.M., Lenstra, J.K., Parrow, J., Woeginger, G.J. (eds.) ICALP 2003. LNCS, vol. 2719, pp. 252–266. Springer, Heidelberg (2003). https://doi.org/10.1007/3-540-45061-0_22
4. Cabac, L., Duvigneau, M., Moldt, D., Rölke, H.: Modeling dynamic architectures using nets-within-nets. In: Ciardo, G., Darondeau, P. (eds.) ICATPN 2005. LNCS, vol. 3536, pp. 148–167. Springer, Heidelberg (2005). https://doi.org/10.1007/11494744_10
5. Camilli, M., Capra, L.: Formal specification and verification of decentralized self-adaptive systems using symmetric nets. Discrete Event Dyn. Syst. **31**(4), 609–657 (2021). https://doi.org/10.1007/s10626-021-00343-3

6. Capra, L.: A Maude implementation of rewritable Petri Nets: a feasible model for dynamically reconfigurable systems. In: Gleirscher, M., van de Pol, J., Woodcock, J. (eds.) Proceedings First Workshop on Applicable Formal Methods, Virtual, 23rd November 2021, Electronic Proceedings in Theoretical Computer Science, vol. 349, pp. 31–49. Open Publishing Association (2021). https://doi.org/10.4204/EPTCS.349.3

7. Capra, L.: Rewriting logic and Petri nets: a natural model for reconfigurable distributed systems. In: Bapi, R., Kulkarni, S., Mohalik, S., Peri, S. (eds.) ICDCIT 2022. LNCS, vol. 13145, pp. 140–156. Springer, Cham (2022). https://doi.org/10.1007/978-3-030-94876-4_9

8. Clavel, M., et al.: All About Maude - A High-Performance Logical Framework: How to Specify, Program, and Verify Systems in Rewriting Logic. Lecture Notes in Computer Science, Springer, Heidelberg (2007). https://doi.org/10.1007/978-3-540-71999-1

9. Ehrig, H., Hoffmann, K., Padberg, J., Prange, U., Ermel, C.: Independence of net transformations and token firing in reconfigurable place/transition systems. In: Kleijn, J., Yakovlev, A. (eds.) ICATPN 2007. LNCS, vol. 4546, pp. 104–123. Springer, Heidelberg (2007). https://doi.org/10.1007/978-3-540-73094-1_9

10. Hoffmann, K., Ehrig, H., Mossakowski, T.: High-level nets with nets and rules as tokens. In: Ciardo, G., Darondeau, P. (eds.) ICATPN 2005. LNCS, vol. 3536, pp. 268–288. Springer, Heidelberg (2005). https://doi.org/10.1007/11494744_16

11. Kahloul, L., Chaoui, A., Djouani, K.: Modeling and analysis of reconfigurable systems using flexible nets. In: Zavoral, F., Yaghob, J., Pichappan, P., El-Qawasmeh, E. (eds.) NDT 2010. CCIS, vol. 87, pp. 343–357. Springer, Heidelberg (2010). https://doi.org/10.1007/978-3-642-14292-5_36. https://doi.org/10.1109/TASE.2010.28

12. Köhler-Bußmeier, M.: Hornets: nets within nets combined with net algebra. In: Franceschinis, G., Wolf, K. (eds.) PETRI NETS 2009. LNCS, vol. 5606, pp. 243–262. Springer, Heidelberg (2009). https://doi.org/10.1007/978-3-642-02424-5_15

13. Llorens, M., Oliver, J.: Structural and dynamic changes in concurrent systems: reconfigurable Petri nets. IEEE Trans. Comput. 53(9), 1147–1158 (2004). https://doi.org/10.1109/TC.2004.66

14. Meseguer, J.: Conditional rewriting logic as a unified model of concurrency. Theor. Comput. Sci. 96(1), 73–155 (1992). https://doi.org/10.1016/0304-3975(92)90182-F

15. Padberg, J., Kahloul, L.: Overview of reconfigurable Petri nets. In: Heckel, R., Taentzer, G. (eds.) Graph Transformation, Specifications, and Nets. LNCS, vol. 10800, pp. 201–222. Springer, Cham (2018). https://doi.org/10.1007/978-3-319-75396-6_11

16. Padberg, J., Schulz, A.: Model checking reconfigurable Petri nets with Maude. In: Echahed, R., Minas, M. (eds.) ICGT 2016. LNCS, vol. 9761, pp. 54–70. Springer, Cham (2016). https://doi.org/10.1007/978-3-319-40530-8_4

17. Prange, U., Ehrig, H., Hoffmann, K., Padberg, J.: Transformations in reconfigurable place/transition systems. In: Degano, P., De Nicola, R., Meseguer, J. (eds.) Concurrency, Graphs and Models. LNCS, vol. 5065, pp. 96–113. Springer, Heidelberg (2008). https://doi.org/10.1007/978-3-540-68679-8_7

18. Reisig, W.: Petri Nets: An Introduction. Springer, New York (1985). https://doi.org/10.1007/978-3-642-69968-9

19. Stehr, M.-O., Meseguer, J., Ölveczky, P.C.: Rewriting logic as a unifying framework for Petri nets. In: Ehrig, H., Padberg, J., Juhás, G., Rozenberg, G. (eds.) Unifying Petri Nets. LNCS, vol. 2128, pp. 250–303. Springer, Heidelberg (2001). https://doi.org/10.1007/3-540-45541-8_9
20. Valk, R.: Object Petri nets. In: Desel, J., Reisig, W., Rozenberg, G. (eds.) ACPN 2003. LNCS, vol. 3098, pp. 819–848. Springer, Heidelberg (2004). https://doi.org/10.1007/978-3-540-27755-2_23

Author Index

Printed in the United States
by Baker & Taylor Publisher Services